SpringerBriefs in Social Policy

SpringerBriefs in Social Policy showcases concise summaries of cutting-edge research and practical applications in a wide variety of current topics in social policy and related areas for a global audience. Featuring compact volumes of 55 to 125 pages, the series covers a range of content from professional to academic. Topics might include:

- A snapshot of a hot and/or emerging topic or development in the field
- A bridge between new research results published in journal articles and a contextual literature review
- A timely report of state-of-the art analytical techniques
- A report from the field
- An in-depth case study
- A presentation of core concepts that students must understand in order to make independent contributions

Both solicited and unsolicited manuscripts are considered for publication in this series.

Briefs are published as part of Springer's eBook collection, with millions of users worldwide, and will also be available as a MyCopy edition. Print editions have a softcover binding.

Briefs have a quick turnaround time and are available for individual print and electronic purchase.

Sana Loue

From Public Policy to Family Dynamics

A Case Study of the Impact of Public Policy on Two 20th Century Jewish Immigrant Families

Springer

Sana Loue
Department of Bioethics, School of Medicine
Case Western Reserve University
Cleveland, OH, USA

ISSN 3059-3026 ISSN 3059-3034 (electronic)
SpringerBriefs in Social Policy
ISBN 978-3-031-71817-5 ISBN 978-3-031-71818-2 (eBook)
https://doi.org/10.1007/978-3-031-71818-2

This Springer imprint is published by the registered company Springer Nature Switzerland AG
The registered company address is: Gewerbestrasse 11, 6330 Cham, Switzerland

Preface

Many books have been written about Jewish immigrants from Eastern Europe during the early twentieth century, the reasons for their emigration from their homelands, their journey to their new countries, and the circumstances in which they found themselves. These works often assume one of two approaches: the provision of a composite picture of the immigrant experience, replete with statistics and documentary evidence, sometimes drawing piecemeal from illustrative experiences of individuals to underscore a particular point or biographies or autobiographies, many of which focus on better educated, more literate, more politically active, or more (in)famous members of the Jewish community or on Holocaust survivors.[1]

In contrast to these approaches, I examine the lives of two Jewish immigrant families at the time of their immigration to the United States, one from what is now Belarus but was once in the Pale of Settlement, and the second from what is now Poland, but also once existed within the Russian Empire. These are my families, that of my paternal and maternal grandparents. By taking this approach, we can better understand how various policies impacted their understandings, perspectives, decisions, and, ultimately, the direction of their lives as well as the lives of those around them.

Each of these chapters examines their lived experience in the context of the larger events occurring around them, over which they had no control. Each chapter refuses to fit comfortably into the master narratives that have emerged over time with regard to Jewish immigration during this period and Jewish families in general: Jewish family cohesiveness, the absence of abuse and neglect within Jewish families, the romanticization of the United States as a safe harbor for Jewish immigrants. Each

[1] For examples of composite portraits, see John Bodnar, *The Transplanted: A History of Immigrants in Urban America* (Bloomington, IN: Indiana University Press, 1985); Oscar Handlin, *The Uprooted: The Epic Story of the Great Migrations That Made the American People*, 2nd ed. (Philadelphia, PA: University of Pennsylvania Press, 2002). Examples of biographies and autobiographies include Michael Shnayerson, *Bugsy Siegel: The Dark Side of the American Dream* (Jewish Lives) (New Haven, CT: Yale University Press, 2021); Jeffrey Rosen, *Louis Brandeis, American Prophet* (New Haven, CT: Yale University Press, 2016).

chapter explores the possible impact, both conscious and unconscious, of existing policies that shaped their experiences and their decisions—the melding of science, medicine, and law to classify and label individuals, to assess individuals' worthiness and value, to shape immigration policy, and to distinguish between those who were worthy and unworthy of economic assistance.

Chapter 1 begins with an account of my youngest brother Michael's birth in 1964 and the response of my father and paternal grandparents to the fact of his Down syndrome. We see his initial separation from his nuclear family, institutionalized in order to alleviate any potential burden on the family. The extent to which our grandparents Leib and Chana, who would become Louis and Annie after their immigration, may have known of or have been affected by the Russian eugenics movement remains unclear, but there is little doubt of their exposure to the core popular ideas of the US eugenics movement relating to the institutionalization and sterilization of those deemed to be "feebleminded" and the need to produce "better babies." They would have read, as well, of the eugenic basis for Hitler's later persecution of Jews. Their son Henry, my father, would have been aware of Nazi efforts to euthanize those whose lives were "not worth living" and, additionally, would have absorbed during his medical training the myth of Down syndrome individuals as nonproductive and essentially nonfunctional. Might Louis and Annie and Henry have carried these experiences with them, even unconsciously, as they projected a future for Michael and fashioned their own responses to his existence?

Chapter 2 begins with my discovery of my mother's passport indicating that she had been born in Poland, contrary to her representations until that time that she had been born in New York City. The story of her immigration unfolds with the immigration of her father Moische, later Morris, claiming status as a single man, despite his then-existing marriage to Gitla, later Gussie, who he had left behind in Poland together with my mother, then Rivka, later to become Rose and then Ruth. We see his bigamous marriage to Dora, the birth of their son Dale, and his later abandonment of them following Gitla's and Rivka's eventual immigration to the United States, in much the same way that he had abandoned Gitla and Rivka following his solo immigration to New York. The chapter explores the possibility that Moische's misrepresentation of his marital status constituted a strategic response to then-existing immigration policy and rising anti-immigrant sentiment by enhancing his desirability as a new immigrant or, perhaps instead, an effort to forge an independent existence for himself.

Chapter 3 brings the past into the present, examining the impact of the various policies on the third generation, my generation, the grandchildren. The chapter explores how policies of the past affected not only those who lived directly under their aegis, but the environments and understandings that were carried forward to later generations as a result.

Because almost all of the recounted events occurred before my birth, much of the story about my family rests on my compilation of available primary documents—ship manifests, marriage and birth certificates, newspaper articles, legislative history, then-current legal and medical literature—and family lore in an effort to understand the impact of policies on my grandparents and their children, the ways in which they navigated society's discourses about and expectations of them, and their decision-making processes. In some ways, my hypothesizing about individuals' underlying emotions, self-perceptions, and motives evokes much-maligned psychohistory. However, my approach departs from psychohistory in that I do not attempt to diagnose; I do not disregard possible childhood or adult influences on the individuals; I do examine the lived context of each individual, and I do not attempt to superimpose a theoretical framework onto the available facts.[2]

William McKinley Runyan, writing about the use of psychohistory as a methodology, noted that an action may have multiple explanations and there may be some truth to each explanation; actions may have multiple causes and meanings. Some explanations may be more valid than others when evaluated against various criteria: they rest on fewer assumptions; are supported by reliable, relevant evidence; can survive efforts to falsify them; are consistent with knowledge about human behavior and/or the specific individual in question; and are credible when evaluated against other explanatory hypotheses.[3] He suggested further that valid psychohistorical writing must reflect three semi-independent levels of inquiry: (1) that which is true of all human beings; (2) that which is true of a specific group of human beings, distinguished by specific characteristics, such as sex, race, ethnicity, or historical period; and (3) that which is true of particular human beings, such as the subject of the writing.[4] In heeding Runyan's note of caution, I both offer multiple possible explanations for individuals' actions to which I was not privy and situate the individuals whose stories I relate in their historical and cultural contexts and in relation to others.

Memory, relying on existing evidence, may be key. Rather than an attempt at the psychoanalysis of individuals as they lived at a time before my birth, I integrate historical evidence with family oral history and my own experiences and conversations with the individuals who appear in this story. As James Wiggins observed in writing about psychohistory:

> Memory is the crux of the matter, I believe. History is sometimes described as an attempt at giving expression to a composite memory, a transcending of the limits of discrete memories as preserved in documents which provide accounts of the event or period under consideration by cross reference to other such documentary memory deposits ... The workings of memory have been better exposited by thinkers familiar with the psychoanalytical process.[5]

---

[2] David Stannard argued that psychohistory suffers from four major shortcomings: therapy, logic, theory, and culture. David E. Stannard, *Shrinking History: On Freud and the Failure of Psychohistory* (New York: Columbia University Press, 1980), 22.

[3] William McKinley Runyan, *Life Histories and Psychobiography: Explorations in Theory and Method* (New York: Oxford University Press, 1984), 43–50.

[4] Runyan, *Life Histories*, 7–9.

[5] James B. Wiggins, "Re-imaging Psycho-History," Theology Today 32, no. 2 (1975): 152–153. Peter Gay observed that both history and psychoanalysis are "sciences of memory." Peter Gay, "Psychoanalysis in History," in *Psychology & Historical Interpretation*, ed. William McKinley Runyan (New York: Oxford University Press, 1988), 107.

You, the reader, might ask why I am willing to self-disclose to this extent, to examine issues that often remain held tightly within family circles, that may be whispered about but never truly investigated or resolved. First, silence allows us to ignore the harmful effects of broadly framed reactionary policies on individuals and families who are captured within their terms. Our averted gaze facilitates our rationalization of both the necessity and the effectiveness of such policies, depriving us of the opportunity to better understand the underlying societal fears, insecurities, and concerns that drove the formulation of such policies in disregard of the human pain they caused and to proactively recognize and intervene in similar circumstances. Second, silence facilitates complicity in the perpetuation of myths: of the United States as a nation characterized by its uniform largesse to and tolerance and acceptance of immigrants; of all Jewish families as close-knit and supportive no matter what the situation; and of all Jews as endowed with great wealth. Third, to remain silent is to participate in the silent and not-so-silent shaming of individuals and families who carry similar stories with them.

The Family Herein

The Family Herein

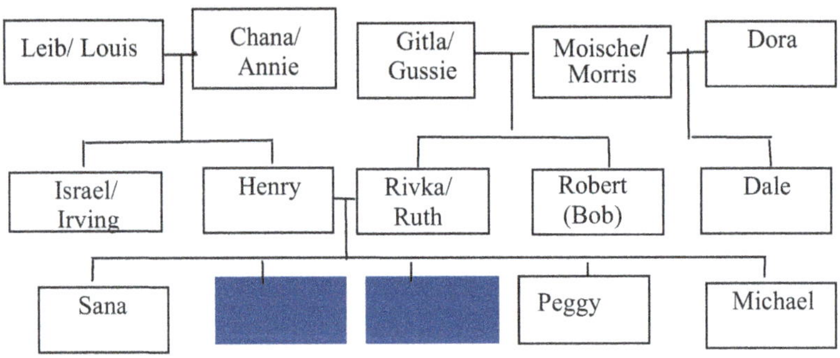

I use the original names of my grandparents—Leib, Chana, Moische, and Gitla—and of my mother Rivka when I refer to events that occurred prior to their arrival in the United States. I use their adopted names—Louis, Annie, Morris, Gussie, and Ruth—in referring to events that occurred after their arrival to the United States. I have not included in the family tree above the names of any cousins, as some requested that they remain unidentified and unidentifiable to the extent possible. Their inclusion in the family tree would render it too easy to unmask their identity.

Cleveland, OH, USA Sana Loue

Acknowledgments

I take full responsibility for both the research and my memories that are reflected in this work. Nevertheless, I could not have completed this manuscript without the support and willingness of others to share their knowledge and understandings with me. Jill Reichenbach of the New York Historical Society guided me in my research of the larger New York City Jewish culture during the relevant periods and American perceptions of Imperial Russia and the Soviet Union. Mia Bruner, Nora Dolliver, and Cathy Sorokurs of the New York Public Library tracked down many of the primary sources that contributed to this work, as did Hallel Yadin of New York-based YIVO Institute for Jewish Research. Dr. Stephen Payne of the Bronx Historical Society spent many hours identifying relevant source materials that form the basis of much of my neighborhood descriptions. Mark Murray, a former immigration officer with what was then the Immigration and Naturalization Service, deciphered cryptic notations on ship manifests that provided clues about my family members' paths to immigration and naturalization in the United States. The insightful feedback of Derek Niemann of the Institute of Continuing Education of the University of Cambridge has helped me to better shape this story. I owe a word of thanks to Arsenii Alenichev for his assistance in tracking down Russian language publications relating to Russian and Soviet eugenics. The incisive questions posed by Elaine Frantz, Mary Ann Heiss, Brian Hayashi, and Sean Martin in response to an earlier version of this work encouraged me to expand my research and my own probing of various issues. I thank the anonymous reviewer of my original proposal for his/her constructive comments. My editor, Janet Kim, has been immensely supportive throughout the process of my writing.

Various organizations should be noted for their willingness to allow me to utilize maps and photographs from their collections. These include New York-based YIVO Institute for Jewish Research, the Ellis Island Foundation, and the New York City Department of City Planning.

I could not have completed this work without having learned to read Russian. The intensive summer Russian language program at Middlebury College deserves a shoutout here for both its rigor and its welcoming environment. I want to thank, in particular, Tom Dolack for his support and his patience with my limitations, and

Holly Myers, for challenging me to engage with Russian language scientific literature.

Copies of many of the family photographs that are included were passed down to me so long ago that I no longer remember from whom they were received. Others were graciously shared with me by my cousin Martin Zand. Special words of appreciation and gratitude are due to my uncle Dale Zand, who spent many hours recounting family history. My sister Peggy Fogelman, eight years my junior, generously provided me with her own insights and perspectives and these are attributed to her where they appear in the text. Several other relatives who trusted me with their knowledge and their memories asked for various reasons that they not be identified within the work. In order to protect their identities, I have not specified in the text my relationship with them, their biological sex, or the date and place of the interview. In all such instances, I have referred to them with a pseudonym. Importantly, I thank my brother Michael for the joy, love, and laughter that we shared throughout his life.

Contents

1 Russian, Soviet, and US Eugenics: Family Internalization of Policy and Rhetoric 1
Leib and the Russian Eugenicists 3
Leib's Immigration Meets US-Style Eugenics 11
Nazi Eugenics 26
References 30

2 The Intertwined Impact of Economics, Eugenic Policy, and Immigration Restrictions 39
Together and Apart: Immigration Policy and the Family 40
The Missing Safety Net: Poverty Unromanticized 55
References 60

3 The Present Past: Policy, Identity, and Progeny 65
Eugenic Policy, Unfulfilled Dreams, and Ambivalent Loss 66
Eugenics, Antisemitism, and "Jewish Self-Hate" 71
Poverty, Shame, Exclusion, and Isolation 78
References 82

Index 87

About the Author

Sana Loue, JD, PhD, MPH, MSSA, MA, LISW-S, CST-T, AVT, is a professor in the Department of Bioethics at Case Western Reserve University School of Medicine in Cleveland, Ohio. She holds secondary appointments in the departments of Psychiatry, Global Health, and Population and Quantitative Health Sciences, as well as in Social Work at the university's School of Social Work. Dr. Loue holds degrees in law (JD), epidemiology (PhD), medical anthropology (PhD), social work (MSSA), public health (MPH), education (MA), theology (MA), and history (MA). Dr. Loue served as the medical school's inaugural Vice Dean for Faculty Development and Diversity from 2012 to 2020.

Dr. Loue has been privileged to conduct research in Laos, Russia, Romania, Tajikistan, and Uganda, and to engage with audiences on issues including the ethical conduct of research, professional development, and diversity in Canada, Chile, Colombia, Kazakhstan, Mexico, Romania, Russia, Spain, Switzerland, Tajikistan, and Vietnam. She has mentored US-based and international graduate students, post-doctoral fellows, and faculty for more than **20 years.**

Dr. Loue's past research in both the domestic and international contexts has focused on HIV risk and prevention, severe mental illness, family violence, and research ethics. Her current research addresses the interplay between religion, society, and bioethics; the integration of cultural humility into clinical care and research settings; and past and current formulations of eugenics. She has authored or edited more than **30** books and has authored/co-authored more than **100** peer-reviewed journal a**rticles.**

List of Figures

Fig. 1.1	Pale of settlement	4
Fig. 1.2	Map of Pogroms in Pale of Settlement, Sites of major pogroms, 1881–1884 and 1903–1906	13
Fig. 1.3	Excerpt of Ship Manifest, SS Latvia, Sailing from Libau, October 18, 1921, arriving at New York City, November 2, 1921	21
Fig. 2.1	Excerpt from Passenger Manifest, S.S. Berengia, Sailing from Cherbourg, France, arriving in New York March 29, 1929	40
Fig. 2.2	Excerpt from Passenger Manifest, S.S. Ryndham, Sailing from Rotterdam, January 25, 1921	42
Fig. 2.3	The Zand Family, circa 1920	43
Fig. 2.4	Moische, Gitla, and Rivka in Lowicz, circa 1920	44
Fig. 2.5	Marketplace in Lowicz	45
Fig. 2.6	Rivka, Poland, unspecified date 1920s	47
Fig. 2.7	Map entitled "New York City (NYC) Maps, Bronx with Community Districts" used with permission of the New York City Department of City Planning. All rights reserved	54
Fig. 2.8	Map entitled "New York City (NYC) Maps, Brooklyn with Community Districts" used with permission of the New York City Department of City Planning. All rights reserved	55
Fig. 2.9	Morris, Gussie, Ruth, and Robert, New York, circa 1937	57

Chapter 1
Russian, Soviet, and US Eugenics: Family Internalization of Policy and Rhetoric

Michael was born on June 25, 1964. He did not come home, but instead my parents drove him to a residential facility, several hours away from where we lived. Born with Down syndrome and a slight heart defect, his prognosis for a normal life span at that time was in doubt.[1] *By that time, my father Henry had been a practicing physician for over a decade and was well aware of the language that had been used during the preceding decade to describe children with Down syndrome: idiot, imbecile, moron, depending on the individual's perceived intellectual development,*[2] *prognostications at birth, without consideration of the environment, whether the child was to be raised in an atmosphere of support and nurturing or merely warehoused.*[3] *These were all words that had been used by the eugenicists to delineate those who were "feebleminded" and undeserving of artificial preservation. The*

[1] There are three types of Down syndrome, but they cannot be distinguished from each other in the absence of chromosomal analysis. Many individuals with Down syndrome also suffer from heart defects, hearing loss, and other chronic health conditions. Down syndrome occurs in 1 of every 700 babies born in the U.S. Centers for Disease Control and Prevention, "Facts about Down Syndrome," April 6, 2021, accessed March 13, 2022, https://www.cdc.gov/ncbddd/birthdefects/downsyndrome.html

[2] President's Committee on Mental Retardation, *Mental Retardation: Past and Present* [MR 76] (Washington, D.C.: Author, Jan. 1977). See also "The Words We Used, 1850s–1940s," Ontario Ministry of Children, Community and Social Services, https://www.mcss.gov.on.ca/en/dshistory/language/1850s_1940s.aspx. The level of intellectual disability among Down syndrome individuals ranges from mild to severe. An individual's ability may be impacted by their environment as well as by the disorder. Center for Parent Information and Resources, "Definition," accessed March 13, 2022, https://www.parentcenterhub.org/downsyndrome/

[3] WillowbrookStateSchool, opened in 1947, serves as a striking example of how children deemed to be mentally retarded were warehoused, often being subjected to physical and emotional abuse as well as experimentation. Although called a school, it was, in fact, a warehouse for handicapped children, with conditions so deplorable that Robert Kennedy referred to it as a "snake pit." John Sibley, "Kennedy Charges Neglect in State Care of Retarded," *New York Times*, September 10, 1965. Willowbrook was not decommissioned until 40 years later, years after a national exposé of

S. Loue, *From Public Policy to Family Dynamics*, SpringerBriefs in Social Policy, https://doi.org/10.1007/978-3-031-71818-2_1

language used to describe Down syndrome children had shifted by the time of Michael's birth.[4] *My parents said he was better off in the institution, that he would be a burden on the family. And, because Michael didn't have a religious circumcision after his birth (a "brit milah"), it was clear that my parents believed or intended that he would never be one of "us," a member of the Jewish community.*[5] *Circumcision represents physically the covenant that exists between God and Abraham mentioned in the Hebrew Bible (Old Testament) and the inclusion of a male in the Jewish faith.*

I visited Michael at the institution once, when I was 11 or 12, and my parents allowed me to accompany them on one of their very occasional visits to see him. I remember very little about that visit, only a large room, with many cribs, the walls gleaming white, Michael in a crib, crying. Years after his birth, maybe when he was two or three years old, my parents had finally agreed to bring home to live with us.

My paternal grandfather Louis, once known as Leib, and my paternal grandmother Annie, previously known as Jennie and Chana even before that, announced excitedly during their visit to us from the Bronx that they saw Michael walk. This was so far from the truth as to be absurd; Michael could barely crawl at that time. They, like my parents, were unable to accept Michael's inborn limitations, demanding with their grimaces, tirades, threats, and sometimes slaps that he conform to their expectations of what a child should be, and yet were also unable or unwilling to contemplate his potential, as if diagnosis were destiny. I wonder now if my grandparents' and parents' response to Michael and his limitations were shaped not only by the disappointment that they may have felt as a result of their unfulfilled expectations of him, but the impact as well of their many years of firsthand and secondhand exposure to the Russian, Nazi, and U.S. eugenic dogma and the societal pressures to bear and raise "wellborn" children.[6]

---

its horrific conditions. See generally James Kazer, A Guide to Willowbrook State School Resources at Other Institutions (Oct. 2005), https://library.csi.cuny.edu/archives/willowbrook-guide

[4] During the 1950s, children were frequently classified according to their perceived functionality: educable, trainable, or custodial. By some time in the 1960s, the language shifted to refer to children with mental retardation as mildly, moderately, or profoundly retarded. "The Words We Used, 1940s–1970s," Ontario Ministry of Children, Community and Social Services, https://www.mcss.gov.on.ca/en/dshistory/language/1940s_1970s.aspx. See also Brenda J. Yumbaco, "State Programs for Retarded: II, Regional Centers and the Future," *The Catholic Transcript* (October 15, 1964): 13.

[5] Circumcision has been characterized as "one of the most enduring rituals" in Judaism. Ohr Margalit and Chariklia Tziraki-Segal, "Circumcision: Man's Obligation and Woman's Praxis," *NASHIM: A Journal of Jewish Women's Studies and Gender Issues* 5767/2006, no. 12 (Fall 5767/2006): 10.

[6] The word eugenics, derived from the Greek word *eugenēs*, meaning "wellborn," was first used in 1883 by Francis Galton. Galton believed that many human physical, mental, and moral traits were inherited, suggesting he need to improve human "stock" through the social control of the generational transmission of various traits. Mark B. Adams, "Eugenics in the History of Science," in The *Wellborn Science: Eugenics in Germany, France, Brazil, and Russia*, edited by Mark B. Adams (New York: Oxford University Press, 1990), 3.

Leib and the Russian Eugenicists

My grandfather Leib likely knew something about the Russian eugenics movement, largely because of its views about Jews, although it would not have affected him directly. Born slightly less than 30 years after Vasily Florinsky's 1865 publication of *Usovershenstvovanie i virzhdeniecheloveshkovoroda* (The Perfection and Degeneration of the Human Race),[7] he left Russia in 1921, just as the eugenics movement was moving toward its heyday in post-Revolution Bolshevik USSR. Little could he know that the immigration restrictions that he would confront in his efforts to immigrate to the USA were occasioned by the American version of eugenics.

Florinsky decried the degeneration of the Russian populace as the result of poorly matched marital partners, diseases, poverty, and slavery. He suggested that humans could be perfected and reach their ideal through an emphasis on good health, wellness, beauty, the achievement of moral excellence, the absence of chronic disease, and the absence of psychopathology, all of which could be achieved through rational marriage effectuated on a population basis. Whether Leib's birth family would have heard of this theory or understood its implications if they had remains unclear. Russian census data for 1897 indicate that less than 50 percent of Russia's male Jews and an even smaller proportion of Russia's female Jews were literate in Russian.[8] Even assuming that those who were literate in Yiddish, the language commonly used among Russia's Jews, were also literate in Russian, they likely would have had more pressing matters than the translation and broadcast in Yiddish of the musings of a Russian scientist that seemed to have little impact on their everyday lives.

For Leib, born on November 6, 1894, lived in the frontier town of Druya, situated about 150 miles east along the Dvina River from Riga and 80 miles or so west of Vitebsk,[9] in the northwest corner of Russia's Pale of Settlement[10] (see Fig. 1.1,

---

[7] Leib was born on November 6, 1894. Find a Grave, accessed December 17, 2020, https://findagrave.com/memorial/150906944/louis-fogelman. Florinsky's publication can be found at Усоверш енствованиеивырождениечеловечскогородаСПБ.: ЖурналДело, 206. For a detailed discussion of Florinsky's work and life and its larger implications, see Nikolai Krementsov, *With and Without Galton: Vasilii Florinskii and the Fate of Eugenics in Russia* (Cambridge, UK: Open Book Publishers, 2018).

[8] Joel Perlman, Literacy among the Jews of Russia in 1897: A Reanalysis of Census Data. [Working Paper 1982]. Jerome Levy Economics Institute, 17, available at https://www.levyinstitute.org/pubs/wp182.pdf

[9] Joe Rapoport, *Echoes of Druya* (Chestnut Ridge, NY: Town House Press, 1989), 3–4.

[10] Most Jews, with few exceptions, were required to live in the Pale of Settlement, established in 1791 by Catherine II until it was abolished in 1917 by a decree of the Provisional Government. "Modern Jewish History: The Pale of Settlement," Virtual Jewish Library, https://www.jewishvirtuallibrary.org/the-pale-of-settlement; "On the Abolition of Religious and National Restrictions," "Об отмене вероисповедных и национальных ограничений". Постановление 20 марта 1917 г. https://web.archive.org/web/20041130033655/ http://humanities.edu.ru/db/msg/41430. The area that was once the Pale now encompasses areas of Belarus, Russia, Ukraine, Latvia, Moldova, and Poland.

Map of Western Russia Showing the Jewish Pale of Settlement.

[The Jewish Encyclopedia, ed. Singer, Funk and Wagner, New York & London, 1901 vol. 10, art. Russia, p. 531]

Fig. 1.1 Pale of settlement. (Herman Rosenthal, J.G. Lipman, Vasili Rosenthal, L. Wygodsky, M. Mysh, Abraham Galante. "Russia," in The Jewish Encyclopedia, vol. 10, Philipson-Samoscz, New York, 1905, 531, Public Domain. Accessed March 23, 2024, https://upload.wikimedia.org/wikipedia/commons/c/c0/Map_showing_the_percentage_of_Jews_in_the_Pale_of_Settlement_and_Congress_Poland%2C_The_Jewish_Encyclopedia_%281905%29.jpg) depicting the Pale of Settlement). Long a part of Tsarist Russia, residents of Druya continually experienced economic hardship, food scarcity, and infectious disease. Leib and his family were not spared; the family had meager means and Leib and a sibling often shared but one bed, despite Leib's frequent complaints about his brother's nightly bedwetting.[11] Druya became part of independent Poland in 1921, following the Russian Revolution and World War I. Those who remained in Druya after the Revolution faced yet again severe hunger and the effects of a typhus epidemic. Although there were no arrests of Jews in Druya and Jews were officially "declassed," they continued to be viewed as enemies of the new state.[12]

Indeed, at the time of Florinsky's publication, the concept of eugenics had not even entered into Russia's medical discourse, let alone popular parlance. By 1904, however, the confluence of a number of factors raised concerns relating to the degeneracy of the Russian population and prompted increased attention to eugenics within medical circles. The word "eugenics" first appeared in 1902 with the release of Liudvik Krzhivitskii's *Psikhicheskierasy* (Psychical Races),[13] a Russian language summary of Galton's views on genetics,[14] followed shortly by Russian translations of the works of a wide range of eugenicists.[15] Sir Francis Galton, a cousin of Charles Darwin,[16] asserted that human traits are both heritable and manipulable for social ends and suggested that a social program of eugenics would "replace natural selection by other processes that are more merciful and not less effective."[17] Between 1904, the Russo-Japanese war, and the onset of the First World War—the Great War as it was and continues to be known in Russia—the country experienced epidemics of cholera, typhus, typhoid, and plague that were both recurrent and severe.[18] Economic hardship prompted worker strikes and peasant uprisings so threatening to

---

[11] Personal communication, Kenneth Fogelman, date unknown.

[12] Rapoport, *Echoes of Druya*, 81–82.

[13] LiudvikKrzhivitskii, Психическиерасы. (St. Petersburg: XX vek), 54–73, 212–223.

[14] Nikolai Krementsov, "The Strength of a Loosely Defined Movement: Eugenics and Medicine in Imperial Russia," *Medical History* 59, no. 1 (2015): 6–31.

[15] N.L. Krementsov, От «звериной философии» к медицинской генетике:
евгеника в России и Советском Союзе, Историко-биологические исследования6, no. 2 (2014): 27. ["From Animal Philosophy to Medical Genetics: Eugenics in Russia and the Soviet Union," *Studies in the History of Biology*].

[16] Charles Darwin was a British naturalist, famous for his theory of evolution. His theory comprised three major elements: that variation occurs randomly within a species, an individual's traits can be inherited by its progeny, and that the struggle for existence would allow only those with the most favorable traits to survive. Charles Darwin, *On the Origin of the Species by Means of Natural Selection, or the Preservation of Favoured Races in the Struggle for Life* (1859), available at https://www.vliz.be/docs/Zeecijfers/Origin_of_Species.pdf

[17] Sir Francis Galton, *Memories of My Life* (London: Methuen), 323.

[18] "Prevention and Control of Epidemics in Russia," УпревлениеделамиПрезидентаРоссийской Федерации [Boris Yeltson Presidential Library], available at https://www.prlib.ru/en/section/1302536. See also Nancy M. Frieden, "The Russian Cholera Epidemic, 1892–1893 and Medical Professionalization," *Journal of Social History* 10, no. 4 (Summer 1977): 538–539; K. David Patterson, "Typhus and Its Control in Russia, 1870–1940," *Medical History* 37 (1993): 361–381. For a discussion of earlier outbreaks, see Arcadius Kahan, "Social Aspects of the Plague

the regime that Tsar Nicholas II issued in October 1905 a manifesto promising civil liberties, the institution of the rule of law, and the creation of a legislative body with elected representatives, all promises that were later abandoned.[19] Although the failed revolution prompted an increase in popularized racist stereotypes of Jews as parasites, revolutionaries, and degenerates, the proliferation of these images was not attributable to eugenicist theories.[20]

The focus on eugenics by Russian scientists and physicians was not, however, borne solely from concern for the state of their country. During these years, the majority of Russian physicians were civil servants, employed in the tsarist court, the army, the navy, or by municipal bodies (*zemstvos*);[21] their employment situation often determined both their self-image and their social status.[22] Medicine's newfound focus on eugenics and the recognition of physicians as experts conferred physicians and scientists with greater stature and the medical profession with increased autonomy from the control of the state.

As the chief inspector in charge of hostels and boarding houses in St. Petersburg, the bacteriologist Nikolai Gamalei became involved in the prevention of epidemic diseases in urban slums. As a resident of a shtetl that was far afield of St. Petersburg, Leib would likely have been unaware of any public health measures that Gamalei might have instituted in an effort to lessen the impact of these epidemics. Gamalei viewed eugenics as an extension of social hygiene applied to human reproduction, rather than a theoretical basis for racial degeneration[23] and disseminated his views with the 1913 publication of *O vyrzdenil: vozrozhdeniiludei po dannyneigenik* (On the degeneration and regeneration of human from a eugenics viewpoint).[24] The

Epidemics in Eighteenth Century Russia," *Economic Development and Cultural Change* 27, no. 2 (Jan. 1979): 255–266.

[19] For a brief history of the 1905 revolution and the tsar's response, see Scott Ury, *Barricades and Banners: The Revolution of 1905 and the Transformation of Warsaw Jewry* (Stanford, CA: Stanford University Press, 2012), 38–44.

[20] Eugene M. Avrutin, "Racial Categories and the Politics of (Jewish) Difference in Late Imperial Russia," *Kritikai: Explorations in Russian and Euroasian History* 8, no. 1 (Winter 2007): 16, 30; Natan Meir, "Jews, Ukrainians, and Russians in Kiev: Interethnic Relations in Late Imperial Associational Life," *Slavic Review* 65, no. 3 (2006): 495–496.

[21] John F. Hutchinson, "Politics and Medical Professionalism After 1905," in *Russia's Missing Middle Class: The Professions in Russian History*, ed. Harley D. Balzer (Armonk, NY: M.E. Sharpe, 1996), 89–116.

[22] Frieden, "Russian Cholera Epidemics," 539. The social status of those entering medicine was often low, as many came from families of pharmacists, barber surgeons, other physicians, soldiers, and low-level bureaucrats. Medicine as a profession was considered to be beneath the status of those with higher social rank. Frieden, "Russian Cholera Epidemic," 540. For an in-depth discussion of the professionalization of Russian medicine and the evolving role and status of physicians, see Nancy Mandelker Frieden, *Russian Physicians in the Era of Reform*, 1856–1905 (Princeton, NJ: Princeton University Press, 1981).

[23] Krementsov, "Loosely Defined Movement," 15.

[24] О выржедении и возрождении лудеи по данным еигеник, Врачебнаиа газета, 7.

journal *Giglena i Sanitariia* (*Hygiene and Sanitation*), which he established in 1910, was the first Russian journal to systematically consider eugenics-related issues.[25]

The psychiatrist Tikhon Iudin was similarly interested in eugenics and, as co-editor of the journal *Soviemennaiapsikhiatriia* (*Modern Psychiatry*), had a platform from which he could expound on his views. In 1907, his study examining the prevalence of psychosis in twins was published in the *Zhurnalnevropatologiii pskhiatrii* (*Journal of Neuropathology and Psychiatry*).[26] Iudin believed that heredity defined the quality of progeny, with only a negligible effect of the environment. However, he characterized the idea of racial superiority and inferiority and policies of sterilization, segregation, and euthanasia, common among his Western eugenicist-counterparts, as both "questionable" and "fanatical."[27] Russian eugenicists not only rejected the participation of physicians in the sterilization of individuals confined to prisons and asylums[28] but also contrary to their Western counterparts, attributed social ills such as crime, alcoholism, prostitution, mental illness, and suicide to political and economic factors.[29] Iudin explained the role of eugenics as both a biotechnical task and a social task in that it relies on the methods of natural science while at the same time, it constitutes a social and historical science.[30] Accordingly, although it was believed that Jews might be at the bottom of the human hierarchy due to their slyness and questionable morality, it was anticipated that those qualities would fade in due time with their civic betterment.[31]

The relative insulation of Leib and his family—his parents and siblings and, later, his wife Chana and son Israel—from any direct effects of the eugenics discourse is not, however, to suggest that they did not otherwise suffer. Although the tsarist regime did not promulgate a racial order premised on biological theories of human development, it promoted race consciousness and racist attitudes.[32] Leib's frequent repudiation of both God and Judaism and his rejection of Jewish rituals would not have shielded him from such attitudes. Current research suggests that prejudice and racism on the one hand and eugenic thinking on the other are, indeed,

---

[25] Krementsov, "Loosely Defined Movement," 10.

[26] Tikhon Iudin, Псикхозы у близнецов. Зхурнал невропатологии и пскхиатрии 7, no. 1 (1907): 68–83.

[27] Krementsov, "Loosely Defined Movement," 20.

[28] Sergei Preobrazhenskil, Кхирургическаиа профилактика вырожденииа. Современнаииа псикхиатрииа 2 (1912): 110–117. [Khirurgicheskaiaprofilaktikavyrozhdeniia; Surgical prevention of degeneration. Sovremennaiiapsikhiatriia (Modern Psychiatry) 2 (1912): 110–117].

[29] Alexander Vucinich, *Darwin in Russian Thought*. Berkeley, CA: University of California Press, 1988.

[30] T.I. Iudin, Eugenics: A Teaching of Improving Natural Qualities in Humans (Moscow: Sabashnikovy, 1925). Quoted in Yu. V. Khen, "Unknown Pages of Russian Eugenics," *Herald of the Russian Academy of Sciences* 76, no. 4 (2006): 386.

[31] See Michael Stanislavski, *Tsar Nicholas I and the Jews: The Transformation of Jewish Society in Russia 1825–1855*. (Philadelphia, PA: Jewish Publication Society of America, 1982).

[32] Avrutin, "Racial Categories," 16.

connected.[33] Antisemitism had longstanding roots of a religious nature;[34] it was later, during the late nineteenth and early twentieth centuries, during the years of Leib's youth and a decade or so beyond his emigration in 1921, that efforts were made to distinguish Jews on the basis of their physical characteristics.[35] Such efforts likely represented, at least in part, a response to the diminution of boundaries between Jews and Christians resulting from name changes, and changes in customs, dress, and religious traditions.[36]

Leib had always asserted that the family name "Fogelman" had been "stolen" and was used in place of the "real" family name of Darsin or Darzin. He had frequently told a tale of having changed the family name after having removed the dog tags from the neck of a dead German soldier, a story that is somewhat plausible in view of Druya's location as a border town near the WWI front. However unlikely, perhaps he had changed the family name through a formal petition process, although such an action might have heightened the potential for conscription. Whether this change of family name was an objective fact or family lore; whether it was to hide the family's Jewish identity, although "Fogelman" is recognizable as a common German Jewish name; whether it was to avoid conscription into the Russian army; or whether it was to somehow facilitate emigration from what had become the Soviet Union; all remains unknown.

With the rise of physical anthropology, scientists believed that they could measure physical features, allowing them as a result to establish the parameters of normalcy and objectively document differences that existed between groups. Aleksei Aren'evich Ivanovskii published in 1903 his construction of a classificatory map of

[33] Eliot Turiel, "Eugenics, Prejudice, and Psychological Research," *Human Development* (2020): doi:10.1159/000512492. Turiel notes that ideas about social characteristics of groups are part of psychological thinking, which is central to eugenic thinking. Eugenic thinking also involves moral judgments.

[34] Neil MacMaster, *Racism in Europe, 1870–2000* (New York: Palgrave, 2001), 104, 107. See also John D. Klier, Shlomo Lambroza, eds. *Pogroms: Anti-Jewish Violence in Modern Russian History* (Cambridge: Cambridge University Press, 1992).

[35] Research indicates that Italian Fascists utilized artwork from different historical periods to promote their ideology. Art works were separated from their original contexts, which allowed the Fascists to superimpose their own meanings on them and articulate their beliefs about race through visual rhetoric. Linday Maldari, "Dismantling an Art History of Eugenics: Fascist Propaganda and Visual Rhetoric in *La difesadellarazza*," Proceedings of the National Conference on Undergraduate Research (NCUR), Kennesaw State University, Kennesaw, Georgia, April 11–13, 2019, accessed April 2, 2024, https://libjournals.unca.edu/ncur/wp-content/uploads/2021/02/2898-Maldari-Lindsay-FINAL.pdf. Improvements in photographic techniques during the late nineteenth century also facilitated the associations between visual representations, race, and character. See Anne Maxwell, *Picture Imperfect: Photography and Eugenics* 1870–1940 (East Sussex, UK: Sussex Academic Press, 2008); David Green, "Veins of resemblance: Photography and Eugenics," *Oxford Art Journal* 7, no. 2 (1984): 3–16.

[36] Harriet Murav, *Identity Theft: The Jew in Imperial Russia and the Case of Avraam Uri Kovner* (Stanford, CA: Stanford University Press, 2001); Eugene M. Avrutin, "A Legible People: Identification Politics and Jewish Accommodation in Tsarist Russia," Ph.D. dissertation, University of Michigan, 2004.

the entire population of the Russian empire, creating a preliminary typology based on 10 indices: the color of the hair and eyes, height, the size of the head, cranial measurements, the length of the face, the nose, the size of the torso, the width of the chest, the size of the arms, and the size of the legs. Despite the deficiencies inherent in his methodology, he concluded that only Jews formed a distinct race.[37]

The 1905 law legalizing confessional transfer led to further invisibility of Jewishness and heightened anxieties relating to the inability to clearly distinguish Jews from their Christian neighbors.[38] Confessional transfer permitted greater religious fluidity, allowing Jews to break from Judaism and Jewishness and be included in Russian society.[39] Because religion became less of a marker of identity, names and language were increasingly utilized to delineate differences between groups. However, restrictions relating to name change were also loosened, so that by the height of the Great War, the Chancellery of Petitions received 2296 name-change petitions, an increase of more than 1500 percent from the 1915 level.

The Russian eugenics movement borrowed extensively and selectively from the components of the USA, German, French, and British eugenics movements; eugenics was a transnational and translingual phenomenon.[40] Russian eugenicists, like their French counterparts, emphasized the potential impact of the environment and education and advocated the improvement of social conditions and the provision of re-education. Most rejected the "negative eugenics" of British, German, and US eugenicists that included involuntary sterilization, segregation and, in the case of the Nazis, killing. These divergences were underscored at the 1912 First International Eugenics Congress held in London. Although Russia did not send official representatives, it was attended by the Russian philosopher and anarchist Petr Kropotkin and the journalist Isaak Schlovskii. Kropotkin attacked the class bias inherent in Western eugenicists' portrayal of workers as unfit.[41] Like Kropotkin, Schlovskii rejected one of the primary underpinnings of the Western eugenics movement, arguing that anthropological research did not support the concept of a racial hierarchy.[42]

Nikolai Konstantinovich Koltsov, who had created the Russian Eugenics Society in 1920, identified three components of eugenics: anthropogenetics, anthropotechnique, and eugenic religion. Anthropogenetics signified the pure science of human heredity. The second component, anthropotechnique, referred to the elucidation of methods that could be utilized to improve the genetic quality of successive

---

[37] Avrutin, "Racial Categories," 33.

[38] Meir, "Jews, Ukrainians, and Russians," 495–496.

[39] Eli Weinerman, "Racism, Racial Prejudice, and Jews in Late Imperial Russia," *Ethnic and Racial Studies* 17, no. 3 (1994): 448, 472.

[40] Krementsov, *With and Without Galton*.

[41] *Problems in Eugenics* (London: Eugenics Education Society, 1913), 50–51. See also Peter Kropotkin, "The Sterilization of the Unfit," *Mother Earth* 7, no. 10 (December 19012). Accessed March 6, 2022, https://sniadecki.wordpress.com/2016/02/04/kropotkine-eugenism-en/

[42] Nikolai Krementsov, "Eugenics in Russia and the Soviet Union," in *The Oxford Handbook of the History of Eugenics*, edited by Alison Bashford and Philippa Levine (New York: Oxford University Press, 2010), 415.

generations. Eugenic religion was analogized to nationalism, as a means to propel the populace to consider self-sacrifice and self-limitations.[43]

The focus of Bolshevik eugenics researchers also differed significantly from that of their Western counterparts. Unlike eugenicists in the USA, e.g., who were preoccupied with the identification of those deemed to be "feeble-minded" or "unfit" and the prevention of their reproduction, Russian eugenicists focused their attention on the heritability of creative abilities, such as those of artists and musicians. The first Russian journal devoted to such research, *the Clinical Archive of Genius-ness and Gifted-ness*, appeared in 1925.[44]

During the early to mid-1920s, Russian eugenicists disseminated their ideas not only to professional audiences through presentations and journal publications but to lay audiences as well, through the dissemination of literature, exhibits, public discussions, and the integration of eugenic principles into the curricula of secondary schools and universities. It may have been as a result of such dissemination efforts that Leib would have learned about Russia's eugenics program. Eugenics even became the focus of popular plays and movies, including собачьесердце,[45] "Heart of a Dog," based on Mikhail Bulgakov's 1925 novel of the same name.[46]

The eugenics movement was ultimately dismissed by 1930, almost a decade after Leib's emigration to the USA. Significant efforts were made following the Revolution to reconcile eugenics theory with the Communist vision. Nevertheless, the movement met its demise, having been characterized as a bourgeoisie science that was incompatible with Bolshevik egalitarianism.[47] Stalin's ascension to power,

[43] Krementsov, "Eugenics in Russia and the Soviet Union," 418.

[44] Irina Sirotkina, *Diagnosing Literary Genius* (Baltimore, MD: Johns Hopkins University Press, 2001).

[45] The film is available at https://russianfilmhub.com

[46] The movie собачьесердце (Sobachieserdtse) is a black-and-white film that follows the plot of the novel. A doctor-scientist transforms the abandoned dog Sharik into a human, named Sharikov, through the implantation of the pituitary gland and testicles of a deceased alcoholic petty criminal. As Sharikov becomes more human, it becomes increasingly evident that he inherited some of the negative traits of his body part-donor, including the use of profanity and heavy drinking. He also retained the hatred of cats that he had had as a dog. After numerous conflicts, the physician-scientist reverses the surgery, and Sharikov is transformed back into the dog Sharik. Banned in the Soviet Union at the time of its production, the film is a critique of both the eugenics movement and the revolutionary movement.

Yvonne Howell, "Eugenics, Rejuvenation, and Bulgakov's Journey into the Heart of Darkness," *Slavic Review* 65, no. 3 (2006): 544–562. As Erica Fudge observed, Bulgakov believed that "just as forced revolution in nature (figured in the transformation of dog into man) creates natural and domestic chaos, so the forced speeding up of change in the political order has likewise created social disorder in Russia." Erica Fudge, "At the Heart of the Home: An Animal Reading of Mikhail Bulgakov's The Heart of a Dog," *Humanimalia: A Journal of Human/Animal Interface Studies* 1, no. 1 (2009). Available at https://strathprints.strath.ac.uk/29516/1/Humanimalia_Bulgakov.pdf

[47] Krementsov, "Eugenics in Russia and the Soviet Union," 420; Loren R. Graham, "Science and Values: The Eugenics Movement in Germany and Russia in the 1920s," *American Historical Review* 82, no. 5 (December 1977): 1144–1145.

the centralization and subordination of science to Marxist-Leninist ideology,[48] the commencement of the Great Terror in 1936, the increasing political tensions between the Soviet Union and Nazi Germany, and the links between fascism and eugenics in Nazi Germany prompted the characterization of Russian eugenicists and geneticists as fascists and led to a number of their executions.[49]

It was during this period that the Russian Eugenics Society established a special commission to study the "Jewish Race." Despite the conclusion of one commissioned study indicating that Jews were not inferior to other groups, the movement continued to portray Jews as a psychologically and physiologically distinct race.[50] There had been indications of increasing hostility in response to Jewish ascendancy in the ranks of the Communist party and participation in the universities, leading to both the depiction of Jewish Bolsheviks without the portrayal of "Jewish features" and the non-acknowledgement of their Jewish background, in an effort to avoid the characterization of the revolution and of all Bolsheviks as Jewish.[51] Whether the pronouncements of the eugenics advocates relating to the distinctiveness of the Jewish race contributed to the hostility against Jews or the efforts of the Party to repress information relating to Bolsheviks' background appears to remain unaddressed.

Leib's Immigration Meets US-Style Eugenics

Neither I nor anyone else now living knows why Leib chose to immigrate to the USA with his wife Chana and infant son Israel, or why he decided to do so when he did. Perhaps it was an attempt to avoid conscription into the Russian army, a fate often dreaded by Russian Jewish families due to its rampant antisemitism and

[48] The Stalinization of science led to the adoption of a neo-Lamarckian view of genetics, which was championed by T. Denisovich Lysenko. Filip Bardziński, "The Concept of the 'New Soviet Man' as a Eugenic Project: Eugenics in Soviet Russia After World War II," *Ethics in Progress* 4, no. 1 (2013): 78. Lysenko argued that acquired traits are subject to inheritance. To believe otherwise would conflict with the Marxist assertion that "human beings are the result of conditions and upbringing." Krementsov, "Beastly Philosophy," 78. For an in-depth discussion of Lysenko's ideas, see Loren Graham, *Lysenko's Ghost: Epigenetics and Russia* (Cambridge, MA: Harvard University Press, 2016). For a comprehensive discussion of post-Revolution eugenics, see Mark B. Adams, "Eugenics in Russia, 1900–1940," in *The Wellborn Science: Eugenics in Germany, France, Brazil, and Russia*, edited by Mark B. Adams (New York: Oxford University Press, 1990), 153–216. A discussion of eugenics and genetics during the Stalinist period can be found in Dmitri Stanchevici, *Stalinist Genetics: The Constitutional Rhetoric of T.D. Lysenko* (Amityville, NY: Baywood Publishing Company, Inc., 2012); Nikolai Krementsov, *Stalinist Science* (Princeton, NJ: Princeton University Press, 1996).

[49] Graham, *Lysenko's Ghost*.

[50] Bardziński, "New Soviet Man," 59. See also Graham, "Science and Values," 1147.

[51] Antony Polonsky, *The Jews in Poland and Russia, vol. III, 1914 to 2008* (Oxford, UK: Littman Library of Jewish Civilization, 2012), 305.

secularization of Jewish men.[52] Maybe Leib or members of his family had suffered through one of more of Russia's many typhus and cholera epidemics and was loathe to confront yet another.[53] Perhaps he got tired of the effort needed to earn sufficient funds to feed his small family or, now that Druya had become part of Poland, feared that he and his family would be targeted in yet another Russian or Polish pogrom, perhaps even worse than the one experienced in Chana's birthplace of Vilna, only a few years prior to their emigration[54] (see Fig. 1.2, depicting sites of pogroms in the Pale of Settlement during select time periods). Leib would have been a young child at the time of the Kishinev pogrom in 1903, and although never in Kishinev, he would have surely heard from the adults around him of the slaughter, rape, and

---

[52] Yohannan Petrovsky-Shtern, *Jews in the Russian Army, 1827–1917* (Cambridge, MA: Cambridge University Press, 1988), 215.

[53] Russia experienced numerous flares of typhus. During the period from 1918 to 1922, it is estimated that up to three million people may have lost their lives to the disease. Patterson, "Typhus and Its Control," 361. Typhus is caused by the rickettsial organism, *Rickettsia prowazeki* and is transmitted by the body louse *Pediculus humanus*. The lice live in people's clothing and attach to humans for blood meals. Symptoms include high fever, nausea, vomiting, mental confusion, and a rash. The disease is associated with poverty and overcrowded living conditions. The Centers for Disease Control and Prevention, "Epidemic Typhus," accessed March 7, 2022, https://www.cdc.gov/typhus/epidemic/index.html. Russia experienced a major cholera epidemic during the years 1892 and 1893, about the time of Leib's birth. For a discussion of this epidemic, see Frieden, "The Russian Cholera Epidemic," 538–559.

[54] Major pogroms in Tsarist Russia occurred in Kishinev (1903); during the period of the failed 1905 Revolution; in the Kovno, Vilna, and Minsk provinces during WWI (1915, 1916); and during the Russian civil war. Jews in Krakow and Vilna in Poland experienced pogroms in 1918 and 1919, respectively. Eugene M. Avrutin and Elissa Bemporad (eds.), *Pogroms: A Documentary History* (New York: Oxford University Press, 2021).

Although a "definitive history" has yet to be written about the pogroms that occurred during Russia's civil war (1918–1920), it has been estimated that more than 1500 pogroms in more than 1300 villages, towns, and cities took place in the Ukraine alone. Oleg Budnitskii, *Russian Jews between the Reds and the Whites, 1917–1920*, translated by Timothy J. Portice (Philadelphia, PA: University of Pennsylvania Press, 2012), 216. Between December 1918 and December 1919, 213 pogroms are attributed to the Whites (17 percent of all pogroms occurring during that interval) and 725 to the Reds (2.3 percent). Budnitskii, *Russian Jews*, 217. Both Pipes and Figes attribute the violence to greed, suggesting that the Jews were perceived to be rich and that the Whites' ideology focused on nationalism. Richard Pipes, *Russia Under the Bolshevik Regime* (New York: Alfred A. Knopf, Inc., 1994), 106; Orlando Figes, *A People's Tragedy: The Russian Revolution, 1891–1924* (New York: Penguin Books, 1998). In contrast, Kerenz maintains that anti-Semitism was a focal point of White ideology that ultimately reached pathological proportion during the civil war. Peter Kerenz, *Civil War in South Russia, 1918–1920* (Berkeley, CA: University of California Press, 1970); Peter Kerenz, "Pogroms and White Ideology in the Russian Civil War," in *Pogroms: Anti-Jewish Violence in Modern Russian History*, ed. John D. Klier and Shlomo Lambroza (Cambridge, UK: Cambridge University Press,1992), 310–311. Budnitskii, in contrast, maintains that anti-Semitism was neither the official doctrine of the Whites nor the policy of the White leadership. Budnitskii, *Russian Jews*, 410. Rather, he views the massive violence as the culmination of anti-Jewish violence that began prior to World War I and Russia's "pathological fear" that the Jews would commit and had committed grievous acts of espionage against Mother Russia. Budnitskii, *Russian Jews*, 225.

Fig. 1.2 Map of Pogroms in Pale of Settlement, Sites of major pogroms, 1881–1884 and 1903–1906. (By permission, from the Archives of The YIVO Institute for Jewish Research, New York)

destruction that befell the Jews,[55] stories that would have been difficult for him to forget.[56] Leib often told a story of having escaped from his incarceration in a Siberian prison, confined as punishment for his participation in the February 1917

[55] The pogrom left 49 Jews killed, 600 raped or wounded, and more than 1000 Jewish-owned homes and shops ransacked and destroyed. For a detailed discussion of the pogrom, which is often said to have marked the beginning of the Nazi campaign against the Jews, see Steven J. Zipperstein, *Pogrom: Kishinev and the Tilt of History* (New York: W.W. Norton & Company, Inc., 2018).

[56] Zipperstein has observed that the Kishinev pogrom "was a moment that cast a shadow so deep, wide, and variegated as to leave its imprint on Jews, on Jew-haters, and on wounds ever since," Zipperstein, *Pogrom*, 23.

revolutionary effort.[57] The Russian revolution and the civil wars were characterized by high levels of anti-Jewish violence and, if Leib had actually participated in the revolutionary effort, he may well have been on the wrong side of somebody.[58] Leib would have been quite young at the time, a mere 22 years old or so. Whether this story was an accurate rendering of events or a tall tale remains unknown to this day.

Leib, like many other Jewish immigrants, may have believed that he had left antisemitism behind when he had left Russia. Such was not the case. Only months before Leib's journey had begun, the US Congress had passed new immigration legislation that severely restricted the ability of Eastern Europeans like Leib to immigrate to the USA,[59] over the strenuous objections of both Jewish and non-Jewish organizations and representatives.[60] Unlike their Russian and Soviet counterparts who advocated "positive eugenics" through re-education of the populace and rational marriage, American eugenicists sought to implement their agenda through "negative" eugenic policies and practices—involuntary sterilization,[61] segregation and anti-miscegenation laws, involuntary institutionalization, and immigration

[57] For a discussion of the circumstances and events leading up to the February 1917 revolution, see Rex A. Wade, *The Russian Revolution, 1917*, third ed. (Cambridge: Cambridge University Press, 2017), 28–52; Mikhail Zygar, *The Empire Must Die: Russia's Revolutionary Collapse, 1900–1917* (New York: Hachette Book Group, 2017), 392–425.

[58] Brendan McGeever, *Antisemitism and the Russian Revolution* (Cambridge: Cambridge University Press, 2019); Jonathan D. Smele, *The "Russian" Civil Wars, 1916–1926* (New York: Oxford University Press, 2017). See also S. Ansky, *The Enemy at His Pleasure: A Journey Through the Jewish Pale of Settlement During World War I*, translated and edited by Joachim Neugroschel (New York: Henry Holt and Company, 2002).

[59] Emergency Immigration Act of 1921, 42 Stat. 5 (May 19, 1921). The formulation of these restrictions rested, at least in part, on a desire to identify and exclude "spies or other dangerous persons" and to "discourage undesirable immigration." A. Warner Parker, "A. Warner Parker on Immigration," *Jewish Immigration Bulletin* (April 1920): 10, 12.

[60] "Jews Denounce Proposed Cut in Immigration," *New York Tribune* (February 14, 1921), 5. The legislation was opposed by various groups that feared it would adversely impact farmers by limiting the numbers of individuals available to work in agriculture. "Senators Find No Emergency for Alien Ban." *New York Tribune*, January 12, 1921, 9. The concern for agricultural labor continues to echo in current legislative debates focused on immigration reform. Janae Francis, "Farmers Lobby for Immigration Reform to Address Labor Shortages," *Christian Science Monitor*, February 23, 2018.

[61] Harry Laughlin of the Eugenics Research Association developed a Model Eugenical Sterilization Law that he presented to the U.S. House of Representatives' Committee on Immigration and Naturalization. His model identified 10 classes of individuals deemed to be "socially inadequate" and deserving of involuntary sterilization: the feeble-minded, the insane, the criminalistic, the epileptic, the inebriate, the diseased, the blind, the deaf, the deformed, and the dependent. The "diseased" included those afflicted with leprosy, tuberculosis, and syphilis. The "dependent" included children and elderly persons who had been confined to "homes," "ne'er-do-wells," tramps, and paupers. The US House Committee on Immigration and Naturalization, *Biological Aspects of Immigration: Hearings*, 16–17 April 1920, 10. A number of states modeled their involuntary sterilizations statutes on Laughlin's model law. Philip R. Reilly, *The Surgical Solution: A History of Involuntary Sterilization in the United States* (Baltimore, MD: Johns Hopkins University Press, 1991).

restrictions, such as those that Leib faced—all in an effort to purify bloodlines and halt the degeneration of the white race.[62]

The US eugenicists borrowed from Darwin's observations relating to the evolutionary survival of the fittest. The physician Charles Davenport advocated reliance on eugenics, "the science of the improvement of the human race by better breeding,"[63] arguing that "[t]he artificial preservation of those whom the operation of natural agencies tends to eliminate … may conceivably destroy the race."[64] Madison Grant similarly voiced concerns regarding the degeneration of the US population, meaning white persons of "good stock" and wealth,[65] due in large part to the infiltration of America by different races.[66]

The eugenicists' link between wealth and fitness reflected earlier economic thought. The economist Simon Patten had theorized:

> Every improvement which simplifies or lessens manual labor, every change in production from a smaller to a larger scale of operations, and every introduction of machinery which displaces skilled labor, increases the amount of deficiencies which the laboring classes may possess without their being thereby overcome in the struggle for subsistence that the survival of the ignorant brings upon society.[67]

Industrialization was, according to Patten, promoting the reproduction of the deficient. The Yale economist Henry Farman had attributed the rise of the unfit to the provision of assistance and called for countervailing measures:

> Every effort that is being made to re-move what Malthus called the "positive checks" to population, without at the same time increasing the preventive checks, must result in an increase of the very classes which are least able to take care of themselves, and render more and more imperative the solution of that exceedingly difficult problem … "the sterilization of the unfit."[68]

Still other economists had blamed the portended demise of "native" stock, meaning Anglo-Saxons, directly on immigration. Francis Amasa Walker opined, "The native element failed to maintain its previous rate of increase because the foreigners

---

[62] Martin S. Pernick, "Eugenics and Public Health in American History," *American Journal of Public Health* 87, no. 11 (1997): 1767–1772; Edward J. Larson, *Sex, Race, and Science: Eugenics in the Deep South* (Baltimore, MD: Johns Hopkins University Press, 1995); Nancy Ordover, *American Eugenics: Race, Queer Anatomy, and the Science of Nationalism* (Minneapolis, MN: University of Minnesota Press, 2003).

[63] Charles Benedict Davenport, *Eugenics. The Science of Human Improvement by Better Breeding* (New York: Henry Holt and Company, 1910).

[64] Charles Davenport, quoted in "Was the Doctor Right?," *Independent*, January 3, 1916, 3.

[65] Grant asserted, "In America we have nearly succeeded in destroying the privilege of birth; that is, the intellectual and moral advantage a man of good stock brings into the world with him. We are now engaged in destroying the privilege of wealth; that is, the reward of successful intelligence and industry …." Madison Grant, *The Passing of the Great Race: The Racial Basis of European History* (New York: Charles Scribner's Sons, 1918), 2–3.

[66] Grant, *Great Race*, 39.

[67] Simon N. Patten, *The Premises of Political Economy: Being a Re-Examination of Certain Fundamental Principles of Economic Science* (New York: Augustus M. Kelley, 1885), 220.

[68] Henry W. Farnam, "The State and the Poor," *Political Science Quarterly* 3, no. 2 (June 1888): 295.

came in such swarms."[69] These new elements, representing "the lowest kind of day labor,"[70] consisted of "immigrants from southern Italy, Hungary, Austria, and Russia," who were "beaten men from beaten races; representing the worst failures in the struggle for existence."[71]

Jews and immigrants were among those who, the "expert" eugenicists thought, should not be afforded artificial preservation.[72] Several years before Leib's arrival, the sociologist Edward Ross described Eastern European Jews as immoral, unethical, and criminal, with an "inborn love of money-making" which they used to achieve "prosperous parasitism,"[73] a professional assessment that mirrored the assertions of the photographer-journalist Jacob Riis several decades earlier.[74] Henry H. Goddard claimed on the basis of intelligence tests administered to newly arriving immigrants to Ellis Island that 83 percent of Jews were "feeble-minded."[75] J. G. Wilson, a physician with the US Public Health Service charged with the responsibility of examining immigrants as they arrived at Ellis Island, had declared, "[T]he Jews are a highly inbred and psychopathically inclined race" as the result of heredity.[76] He countered Jews' disagreement with his assessment with the claim that the "general paranoid tendency of the race is shown in an almost universal tendency … to fail to appreciate the point of view of the one who opposes them."[77]

[69] Francis Amasa Walker, *Discussions in Economics and Statistics*, vol. 2, edited by Davis Dewey (New York: Henry Holt and Co., 1899), 423.

[70] Walker, "Discussions," 424.

[71] Francis Amasa Walker, "Restriction of Immigration," *Atlantic* 77 (1896): 828.

[72] The catalog of those thought to be undeserving of "artificial preservation" was a long one that included, in addition to Jews and immigrants, women, the working poor, persons of color, criminals, persons deemed to be feebleminded, sick and crippled persons, and those who were alleged to be "morally deficient." John R. Commons had argued that slavery was a necessity because "the negro could not possibly have found a place in American industry had he come as a free man … if such races are to adopt that industrious life which is a second nature to races of the temperate zones, it is only through some form of compulsion." John R. Commons, *Races and Immigrants in America* (New York: Chautauqua Press, 1907), 136. Henry Rogers Seager had argued that individuals with disabilities should be isolated or sterilized in order to "maintain a race that is to be made up of capable, efficient and independent individuals and family groups." Henry Rogers Seager, "The Minimum Wage as Part of a Program for Social Reform," *Annals of the American Academy of Political and Social Science* 48 (July 1913): 10.

[73] Edward A. Ross, "The Hebrews of Eastern Europe in America," *The Century Magazine* 88 (Sept. 1914): 785–792.

[74] Jacob Riis, *How the Other Half Lives: Studies among the Tenements of New York* (New York: Penguin Books, 1997 [1890]). Riis characterized Jews as inherently greedy and exploitative of Christians.

[75] Henry H. Goddard, "Mental Tests and the Immigrant," *Journal of Delinquency* 2, no. 5 (September 1917): 244, 252.

[76] J.G. Wilson, "A Study in Jewish Psychopathology," *Popular Science Monthly* 82 (1913): 265. Wilson went on to suggest that, because Jews generally married other Jews, they were guilty of racial incest.

[77] Wilson, "Jewish Psychopathology," 271. Jews were publicly criticized "for lack of sincerity" in responding to situations in which they believed they were treated unfairly due to their religion. "Asks Jews to Meet All Attacks Fairly," *New York Times* (June 19, 1922), 9.

Wilson was apparently either ignorant or dismissive of actual data; a contemporaneously published report as well as a later study that both focused on the incidence and prevalence of mental illness among Jews in New York during this period of time reported from then-available data that Jews actually accounted for a lower proportion of admissions to psychiatric hospitals relative to their proportion of the population, as compared to non-Jews.[78]

Nevertheless, Wilson's views echoed those of the psychiatric professionals of the day, who often associated specific mental illnesses with specific groups of immigrants, frequently in the absence of any empirical data to support their assertions: manic depressive psychosis, dementia praecox [schizophrenia], psychoneuroses, psychoses, and "hidden sexual complexes" with Hebrews; suicide and other self-destructive behaviors with Japanese; persecutory delusions with West Indians; and alcoholic psychoses with Slavs.[79] It is unclear whether these diagnoses were correct,

[78] Jacob A. Goldberg, *Incidence of Insanity among Jews* (New York: National Committee for Mental Hygiene, 1922); Benjamin Malzberg, "Mental Diseases among Jews in New York State, 1920 to 1952," *YIVO Annual of Jewish Social Science* 10 (1955): 279–199. Bernheimer, however, claimed that "[i]nsanity is very frequent among the Jews" and that "the Jews in [New York City] supply a greater number of insane to the asylums than any other race living here." Charles Seligman Bernheimer, *The Russian Jews in the United States: Studies of Social Conditions in New York, Philadelphia and Chicago, with a Description of Rural Settlements* (Andesite Press, 1905), 295. He did not refer to any particular study in making this claim. The ostensibly greater proclivity of Jews to mental illness was explained as "the outcome of a long series of events in the Jews' history for the last two thousand years. It is a result of the anxiety, prolonged worry, and cerebral overwork of the Jews under the ban of medieval persecution." Bernheimer, *The Russian Jew*, 295. Howe repeated this assertion, claiming that Jews were particularly susceptible to neurasthenia, hysteria, and breakdown due to overwork, dislocation, and "an inability to adapt to new circumstances." Irving Howe, *World of Our Fathers: The Journey of East European Jews to America and the Life They Found and Made* (New York: New York University Press, 2005),149. No evidence was cited for this claim.

Whether or not it was true of the Jews at that time, the attribution of mental illness to historical, intergenerational events is now recognized as intergenerational historical trauma, which has been the focus of research with succeeding generations of Holocaust survivors, Native Americans, and descendants of the US slaves. Amy Bombay, Kimberly Matheson, and Hymie Anisman, "The Intergenerational Effects of Indian Residential Schools: Implications for the Concept of Historical Trauma," *Transcultural Psychiatry* 51, no. 3 (2014): 320–338; Rachel Lev-Wiesel, "Intergenerational Transmission of Trauma Across the Generations: A Preliminary Study," *Quantitative Social Work* 6, no. 1 (2007): 75–94; Hadas Wiseman, Einat Metzl, and Jacques P. Barber, "Anger, Guilt, and Intergenerational Communication of Trauma in the Interpersonal Narratives of Second Generation Holocaust Survivors," *American Journal of Orthopsychiatry* 76, no. 2 (2006): 176–184; Angela Connolly, "Healing the Wounds of Our Fathers: Intergenerational Trauma, Memory, Symbolization and Narrative," *Journal of Analytical Psychology* 56 (2011): 607–626. For a discussion of the mechanisms by which trauma is transmitted and the continuity of self, see Doris Brothers, "Traumatic Attachments: Intergenerational Trauma, Dissociation, and the Analytic Relationship," *International Journal of Psychoanalytic Self Psychology* 9, no. 1 (2014): 3–15; Yael Danieli, "Assessing Trauma Across Cultures From a Multigenerational Perspective," in *Cross-Cultural Assessment of Psychological Trauma and PTSD*, ed. John P. Wilson and Catherin So-kum Tang (Boston, MA: Springer, 2007), 65–89.

[79] James V. May, *Mental Diseases: A Public Health Problem* (Boston, MA: Richard G. Badger, 1922), 160–161. Bernheimer also suggested that Jews were more likely than other groups to suffer from neurasthenia, hysteria, and anxiety. Bernheimer, *The Russian Jew*, 295.

even given differences in diagnostic assessments between then and now, or whether the diagnoses resulted, at least in part, from a misinterpretation or misunderstanding of the cultural significance of specific emotional and/or behavioral responses to stress.[80] And, while the diagnosis of a specific individual may have been correct, the generalization of a diagnosis to a population group would not have been.

Too, at a time when both social mores and legislative acts demanded the separation of blacks and whites, and science claimed the ability to distinguish between the races based on their physiognomy, the whiteness of Jews was called into question.[81] The massive influx of Eastern Europeans to the eastern cities and of Chinese immigrants to California prompted fears of contagion, both literally and figuratively;[82] immigrants were equated with a "diseased organism."[83] Indeed, only 20 years or so before Leib's entry into the USA, Jews in New York City had been subjected to quarantine, blamed for the typhus and cholera epidemics that swept the city in 1892.[84] While the infections would pass, the stigma that had been associated with the disease and with Jews as a group remained. The characterization of immigrants as "foreign colonies embedded in the fair physical corpus of New York"[85] almost heralded the Nazis' later portrayal of the Jews as a cancer in the Aryan body, necessitating their excision.[86] Burton J. Hendrick portrayed Jews as an invading non-white horde:

---

[80] Psychiatric misdiagnoses due to such misinterpretations appear to occur more frequently when assessments are cross-cultural in context. For example, Puerto Ricans experiencing *ataque de nervios*, literally a nervous attack, may be suffering from symptoms of a mental illness, or may be simply communicating extreme stress; a determination requires a deeper inquiry and cannot be made on the basis of observed behavior alone. Sana Loue, *"My Nerves Are Bad": Puerto Rican Women Managing Mental Illness and HIV Risk* (Nashville, TN: Vanderbilt University Press, 2011), 20–24; Peter J. Guarnaccia, "Ataques de Nervios in Puerto Rico: Culture-Bound Syndrome or Popular Illness?" *Medical Anthropology* 15 (1993): 157–270; Peter J. Guarnaccia, Roberto Lewis-Fernandez, and Melissa Rivera Marano, "Toward a Puerto Rican Popular Nosology: Nervios and Ataques de Nervios." *Culture, Medicine, and Psychiatry* 27 (2003): 339–366.

[81] Karen Bodkin, *How Jews Became White Folks and What That Says about Race in America* (New Brunswick, NJ: Rutgers University Press, 1990); Matthew Frye Jacobson, *Whiteness of a Different Color: European Immigrants and the Alchemy of Race* (Cambridge, MA: Harvard University Press, 1998).

[82] One writer had asserted that "the law-abiding citizen fears from the immigrant not only the germ of bodily disease, but the germ of anarchy and also favorable media for its growth." A.J. McLaughlin, "The American's Distrust of the Immigrant," *Popular Science Monthly* 62 (1903): 231.

[83] Gerald V. O'Brien, "Indigestible Food, Conquering Hordes, and Waste Materials: Metaphors of Immigrants and the Early Immigration Restriction Debate in the United States," *Metaphor and Symbol* 18, no. 1 (2003): 35–38.

[84] Howard Markel, *Quarantine! East European Jewish Immigrants and the New York City Epidemics of 1892* (Baltimore, MD: Johns Hopkins University Press, 1997).

[85] E.F. Frazer, "Our Foreign Cities: New York," *The Saturday Evening Post* 1953, no. 1 (June 16, 1923): 6.

[86] Alfred Rosenberg, *The Myth of the Twentieth Century* (CreateSpace Independent Publishing Platform, 2011).

> Thus the blood of this Turkish or Mongol people flows extensively in the veins of the Eastern Jew of today. A further large Slavic mixture makes the Eastern Jew racially alien to Jews from other parts of Europe. Thus the masses that comprise one fourth the present population of New York City trace their beginnings, in considerable degree, to certain tribes that roamed the steppes of Russia in the Middle Ages and happened to accept the religion of Judah as their own.[87]

The lawyer T. Lothrop Stoddard, a staunch eugenicist, attributed "the frizzy or woolly hair, thick lips, and prognathous jaws appearing in many Jewish individuals" to the "Negroid strain" among Jews, further supposed evidence of Jews' racial inferiority.[88]

The Jews, then, were not only different, but a race apart, and an inferior and defective one at that, short of stature with a poorly developed muscular system, frequent anemia, in all, having the appearance of "a sickly people."[89] This view, as well as the popular, albeit false, attribution of crime to the immigrants, underlay the later enactment by the US Congress of additional ethnic restrictions on immigration.[90]

Years before the promulgation and passage of the restrictionist immigration laws, Prescott Hall, a vocal opponent of unrestricted immigration, had warned that "the physical degeneration of the Jew in New York and Philadelphia has been accompanied to some extent by a moral and political degeneration."[91] Madison Grant, once called the "high Priest of racialism in America,"[92] condemned "the Jew, whose dwarf stature, peculiar mentality, and ruthless concentration on self-interest are being engrafted upon the stock of the nation."[93]

Proponents of a eugenics agenda wielded both significant social and political power. They conceived of the "immigration problem" as a racial problem. Robert De Courcy Ward, a Harvard University professor of climatology, argued:

---

[87] Burton J. Hendrick, *The Jews in America* (Garden City, NY: Doubleday, Page and Co., 1923), 96.

[88] T. Lothrop Stoddard, "The Pedigree of Judah," *Forum* (March 1926): 326.

[89] Bernheimer, *The Russian Jew*, 282–283.

[90] With respect to the relationship between the restrictions and the perception of Jews as sickly or inferior, see Michael Rogin, *Blackface, White Noise: Jewish Immigrants in the Hollywood Melting Pot* (Berkeley, CA: University of California Press, 1996), 101; Paula S. Rothenberg, *Race, Class, and Gender in the United States: A Integrated Study*, seventh ed. (New York: Worth Publishers, 2007), 99. For a discussion of the purported association between the immigrants and crime, see The attribution of high rates of crime to the immigrants, which continued long after the passage of the 1924 restrictions, was vehemently contested in the Jewish press. E.g., "U.S. Census Figures Refute Johnson's Accusation of Crime Among Immigrants," *Jewish Daily Bulletin*, November 11, 1927, 1, 3; "Brith C.C. Van Vechton, "The Criminality of the Foreign Born," *Journal of Criminal Law and Criminology* 32, no. 2 (1941–1942): 139–147. Shalom and B'nai Brith Representatives Score Linking of Immigrants with Crime," *Jewish Daily Bulletin*, September 6, 1932, 3.

[91] Prescott F. Hall, *Immigration and Its Effects upon the United States*, second rev. ed. (New York: Henry Holt and Company, 1908), 51–52.

[92] Gunnar Myrdal, *American Dilemma: The Negro Problem and Modern Democracy* (New York: Harper & Brothers, 1944), 14.

[93] Grant, *Great Race*, 14.

> We are allowing race mixing to proceed on a large scale without definitively knowing, for instance, which characteristics of the white, the yellow, and the black breeds are dominant; without having a knowledge of whether certain undesirable physical and psychic characters of other races are not so strongly dominant as to breed out some of the most desirable and distinctive characteristics of the aryo-germanic race, which took possession of this continent, assumed domination over it and planted in it its civilization, institutions, and ideals, [but] is … in danger of being mixed with the blood of other races to the degree of obliteration if the present influx of immigration from the far east, from southern and southeastern Europe and from Asia Minor is allowed to go unchecked …. The final test, which is inevitably approaching, will not consist in assimilating this or that other race but in dominating the complicated network of heredity in what threatens to become a blood chaos.[94]

Leib, Chana, and Israel arrived in the USA on November 2, 1921, during the heyday of the American eugenics movement[95] (Fig. 1.3 provides an excerpt of the ship manifest containing the names of Leib, Chana, and Israel). Leib may not have known about or understood the basis of these new restrictions, but he had somehow learned that they existed and that he needed to circumvent them.[96] Exceptions were made in few circumstances, and Leib did his best to fall within these exceptions. His older brother Samuel had become a USA citizen, conferring a status on Leib that would enable him to bypass the quota restrictions. However, Leib left nothing to chance; he represented to the shipping agent that he was an electrotechnician, a skill that immediately placed him in a category of skilled workers, entitling him to a special exemption.[97] Leib, though, had never been an electrotechnician; he was, in fact, a house painter and would continue with that work after his entry into the USA.

The eugenic rhetoric continued, assuming various forms and seemingly pervading all aspects of daily living. Rudolph Pintner, a professor at Teachers College of Columbia University, asserted that "the races from the south and east of Europe seem inferior in intelligence to those from the north and west."[98] The educator and eugenicist Harry Laughlin remained dissatisfied with the decreased numbers of immigrants from Eastern Europe following the promulgation of the 1921 immigration laws and that were further reduced as a result of additional restrictions that became effective in 1924. Even during the Nazi onslaught against the Jews, Laughlin warned in 1939 that the USA would suffer "conquest by settlement and

---

[94] Robert De Courcy Ward, "Race Genetics Problems," *American Breeders' Magazine* 2, no. 3 (Third Quarter 1911): 232.

[95] Pernick, "Eugenics and Public Health," 1767.

[96] Although Leib and Chana would have faced fewer restrictions *entering into the U.S.* had they immigrated earlier, they likely would have faced significant hurdles *leaving Russia* if they had attempted to leave prior to the Russian Revolution. For a description of the emigration challenges under tsarist rule, see Gur Alroey, "Out of the Shetl. In the Footsteps of Eastern European Jewish Emigrants to America, 1900–1914," *Leidschrift: Organisatie En Regulering Van Migratie in Di NieuweTijd*22, no. 1 (April 2007): 91–122.

[97] Personal communication, Mark Murray, former examiner with the Immigration and Naturalization Service, May 12, 2021.

[98] Rudolph Pintner, *Intelligence Testing: Methods and Results* (New York: Henry Holt, 1923), 361.

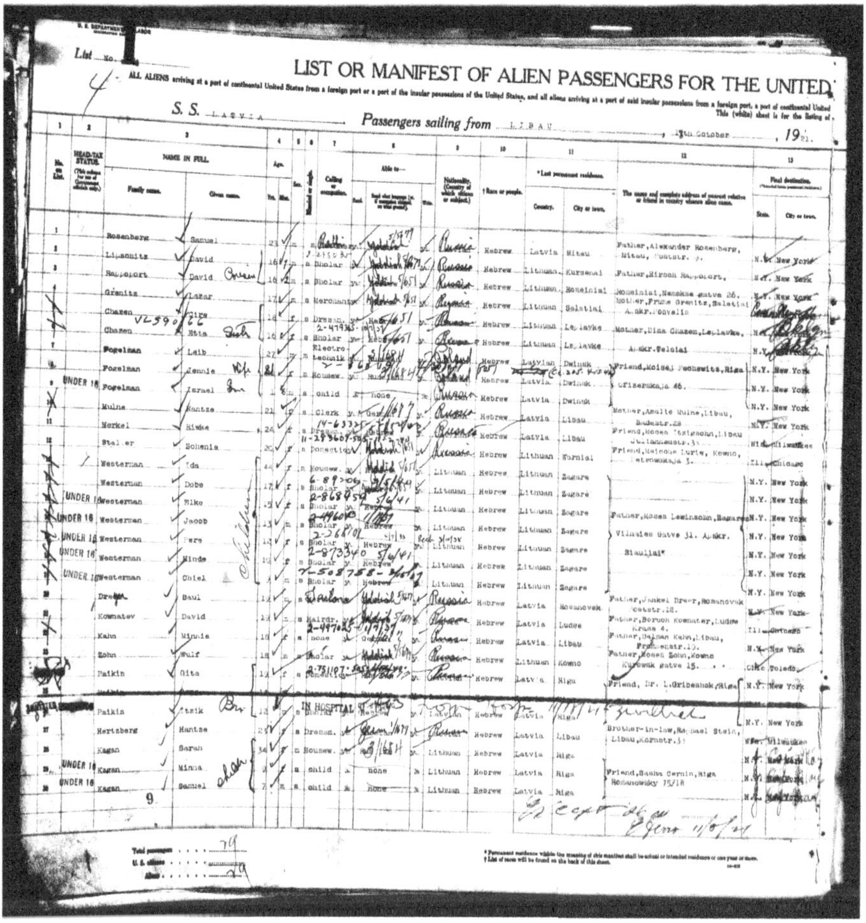
LIST OR MANIFEST OF ALIEN PASSENGERS FOR THE UNITED

S. S. LATVIA Passengers sailing from LIBAU, 18th October, 1921.

Fig. 1.3 Excerpt of Ship Manifest, SS Latvia, Sailing from Libau, October 18, 1921, arriving at New York City, November 2, 1921. (Source: https://heritage.statueofliberty.org/)

reproduction" due to infestation by defective immigrants.[99] C. G. Campbell, then the president of the Eugenics Research Association, lauded the Nazis' Nuremberg Laws that defined Jews as a separate race and forbade sexual relations and intermarriages between Jews and those with German or related blood,[100] claiming that "No

[99] Harry H. Laughlin, *Immigration and Conquest: A Report of the Special Committee on Immigration and Naturalization of the Chamber of Commerce of the State of New York* (New York: Special Committee on Immigration and Naturalization of the Chamber of Commerce of the State of New York, 1939), i, 8, 9.

[100] The Nuremberg Laws, announced by the Nazi regime on September 15, 1935, consisted of two laws: the Reich Citizenship Law, which specified that only racially pure Germans could hold German citizenship, and the Law for the Protection of German Blood and German Honor, which banned sexual relations and intermarriage with Jews. "The Nuremberg Race Laws," *Holocaust*

earnest eugenicist can fail to give approbation to such a national policy. Indeed it goes to realize the hopes that eugenicists have entertained for many years, but have despaired of ever seeing adopted in the present generation."[101]

And Jews, in many ways, were to be kept a race apart. While not approaching the severity of the economic and social restrictions that he had likely endured in Russia, Leib was to confront America's own brand of discrimination, often premised on eugenic beliefs about the Jewish "race." Although many Jews would later ascribe the discrimination that they suffered in America as stemming from religious bigotry, the reality was that it affected Jews, secularist and observant alike.[102] Advertisements for boarders specified "Jews need not apply."[103] Jews were excluded from clubs and associations, and preeminent universities established quotas limiting the number of Jews who could attend.[104]

It was during this time, as well, that the English language had become a symbol of national unity and identity, yet another means by which to discriminate between those seen as racially, intellectually, and morally inferior and those who were considered to be Americans.[105] "[D]ifferentials in literacy [were] tied to national origin and race, in the context of eugenics."[106] Between 1917 and 1922, various states implemented legislation that required immigrants to attend public evening schools if they were unable to speak or read English, and 34 had passed English-only laws.[107]

One writer from the educational establishment went so far as to claim that American students would be contaminated by speakers of other languages, including Russian, and therefore required protection.[108] Not surprisingly, the children of immigrants often developed feelings of shame and inferiority in response to society's

Encyclopedia, US Holocaust Memorial Museum, accessed March 6, 2022, https://encyclopedia.ushmm.org/content/en/article/the-nuremberg-race-laws

101 C.G. Campbell, "The German Racial Policy," *Eugenical News* 21, no. 2 (1936): 25. See also "U.S. Eugenist Hails Nazi Racial Policy," *New York Times*, August 29, 1935. Referring to Nazi racial policies, Campbell asserted, "It sets a pattern which other nations and racial groups must follow if they do not wish to fall behind in their racial quality, in their racial accomplishments and in their prospects for survival."

102 Naomi W. Cohen, "Antisemitism in the Gilded Age: The Jewish View," *Jewish Social Studies* 41, no. 3/4 (Summer-Autumn 1979): 187–188.

103 Cohen, "Gilded Age," 188; "Dr. Rabinow Deplores Bias Against Jew in America," *Jewish Daily Bulletin*, July 6, 1930, 2.

104 Leonard Dinnerstein, *Antisemitism in America* (New York: Oxford University Press, Inc., 1994), 87. See, for example, "Y.M.C.A. Bars Vote to Jew," *New York Herald*, February 20, 1921, 16; "To Discuss Jew in College," *New York Times* (December 23, 1922), 15.

105 Aneta Pavlenko, "'We Have Room for But One Language Here': Language and National Identity in the US at the Turn of the twentieth Century," *Multilingua* 21 (2002): 175.

106 Steven Selden, "Transforming Better Babies into Fitter Families" Archival Resources and the History of the American Eugenics Movement, 1908–1930." *Proceedings of the American Philosophical Society* 149, no. 2 (June 2005): 214.

107 Edward George Hartmann, *The Movement to Americanize the Immigrant* (New York: Columbia University Press, 1948).

108 Miles H. Gordy, "The German Language in Our Schools," *Educational Review*, October (1918): 257–263.

demand for perfect, accent-less English as the price for belonging.[109] Perhaps it was in response to such expectations that Jewish poets who immigrated from Europe in the late 1800s and early 1900s wrote in Yiddish, suggesting a desire to become American with respect to the content of their work but, by writing in a language that was inaccessible to most Americans, evidencing their refusal to become yet another indistinguishable element within the larger "melting pot."[110]

Jewish fiction writers during this period of time may have failed to appreciate the potential ramifications of their portrayals of Jews as a separate, identifiable race. The fictional works often portrayed Jews as "a racial unity, a unity which they described as a nonrational, uncontrollable force that separated Jews from mainstream society."[111] This depiction, however, likely worked to their disadvantage during this era. While it helped to explain the community-wide bonds that existed across diverse segments of the Jewish community, it reinforced the assertion that Jews were of divided loyalties and therefore could not be full-fledged and equal citizens.[112]

Louis and Annie, as Leib and Chana became known in the USA, were likely reminded on a relatively frequent basis of society's fixation with the production of wellborn children. Throughout the 1920s and 1930s, major magazines, including

---

[109] Stephen Brumberg, *Going to America, Going to School: The Jewish Immigrant Public School Encounter in Turn-of-the-Century New York City* (New York: Praeger, 1986); Alfred Kazin, *A Walker in the City* (New York: Harcourt, Brace, and World, 1951).

[110] Cristina Stanciu, "Strangers in America: Yiddish Poetry at the Turn of the Twentieth Century and the Demands of Americanization," *College English* 76, no. 1 (September 2013): 61.

[111] Nadia Malinovich, "Race and the Construction of Jewish Identity in French and American Jewish Fiction of the 1920s," *Jewish History* 19, no. 1 (2005): 29.

[112] The belief that Jews were of divided loyalties was carried to its extreme by Henry Ford's republication of "The Protocols of the Learned Elders of Zion," which was purported to have been authored by Jews as essentially a manifesto to control the global economy. Portions of the "Protocol" appeared the *Dearborn Independent*, a newspaper owned by Ford. Dearborn Independent, *The International Jew: The World's Foremost Problem: Being a Reprint of a Series of Articles Appearing in the Dearborn Independent from May 22 to October 2, 1920.* (Forgotten Books). Ford later published in 1927 "The Scope of Jewish Dictatorship in the U.S.," in which he accused Jewish bankers and merchants of a conspiracy to control U.S. wheat farms. Gary Phillip Zola and Marc Dollinger, eds., *American Jewish History: A Primary Source Reader* (Waltham, MA: Brandeis University Press, 2014), 217. The dissemination of the document is traced in "Story on How the 'Protocols' Were Spread Here and Abroad," *New York Times*, September 4, 1921, 81. Aaron Sapiro, a Jewish agricultural organizer, sued Ford for defamation after being attacked by name in these articles. Ford stopped publishing anti-Jewish materials and issued public apologies to individuals and groups, which may have been part of the out-of-court settlement reached with Sapiro. Isaac Landeman, then-editor of the *American Hebrew*, had challenged in 1920 Ford's assertion of a conspiracy to "establish a Jewish imperialism" and had demanded a public apology. "Challenges Ford to Jewish Inquiry," *New York Times*, November 12, 1920, 5.

The status of Jews as Jews or Americans or both was a frequent focus of discussion within the Jewish community itself. For example, the well-known Rabbi Stephen Wise devoted talks to this issue. *New York Herald*, December 15, 1922, 12. In writing about Americanization and Africanist characters and perspectives, Toni Morrison observed, "Deep within the word 'American' is its association with race." Toni Morrison, *Playing in the Dark: Whiteness and the Literary Imagination* (New York: Vintage-Random House, 1993), 47.

Good Housekeeping, *Saturday Evening Post*, and *Popular Science Monthly* carried articles touting the benefits of eugenic "science."[113] Newspapers often contained notices of upcoming "better baby contests,"[114] and it would have been difficult for Louis to avoid seeing their appearance in his favored newspaper, *The Daily News*. Announcements of these contests continued at least through the late 1940s, a time when Louis and Annie's son—my father, Henry—then in his mid-twenties, surely would have been aware of them. These contests had winners and losers and, consequently, served to demonstrate who should and should not reproduce and whose life was to be deemed suitable.[115] This messaging was later reinforced even more strongly with the evolution of the better baby contests into "fitter family contests," essentially "draw[ing] on and reinforce[ing] a specifically Darwinian equation of competition with health"[116] and the consequent need to control human reproduction and heredity.[117]

Perhaps my grandparents and father absorbed these messages and eugenicists' continuing attacks on public health and social welfare programs that aided sick people long enough to enable them to reproduce and pass on their hereditary defects to subsequent generations.[118] This attribution of child characteristics to heredity "meant holding the parents morally responsible for having caused it."[119] Ruth, my mother, may well have felt some guilt for having produced a "defective" child; this was an era during which mothers were blamed for all of the faults and deficiencies of their offspring.[120] And my grandparents would have likely read in a newspaper or heard on the radio about the Supreme Court decision in the case of *Buck v. Bell*,[121] in which the Court upheld the constitutionality of Virginia's involuntary sterilization law. The Court's decision allowed the state to proceed with the involuntary sterilization of Carrie Buck, alleged to have been mentally deficient and a moral delinquent after becoming pregnant.[122] In rendering the decision of the Court, Justice Holmes

---

[113] Garland E. Allen, "The Social and Economic Origins of Genetic Determinism: A Case History of the American Eugenics Movement, 1900–1940 and Its Lessons for Today," *Genetica* 99 (1997): 81.

[114] See, e.g., "Ashland Branch 'Y' to Hold Baby Contest in Connection with Fall Fair," *The New York Age*, October 18, 1924; "Ashland Place Branch 'Y' to Hear Miss Eva Bowles; Babies Who Won Awards," *The New York Age*, November 8, 1924; "Better Baby Contest," *Daily News*, May 12, 1949.

[115] Martin S. Pernick, "Taking the Better Baby Contests Seriously," *American Journal of Public Health* 92, no. 5 (May 2002): 707.

[116] Pernick, "Better Baby Contests," 708.

[117] Selden, "Transforming Better Babies," 221.

[118] See Pernick, "Better Baby Contests," for the denunciation of public health and social welfare programs.

[119] Pernick, "Eugenics and Public Health," 1769.

[120] Ruth Rosen, *The World Split Open: How the Modern Women's Movement Changed America* (New York: Penguin Books, 2000), 15.

[121] *Buck v. Bell*, 274 U.S. 200 (1927).

[122] Carrie Buck was committed to the Virginia Sate Colony for Epileptics and feeble-minded by her foster parents after they noted that their 17-year-old ward was pregnant. Carrie was the daughter of

opined, "Instead of waiting to execute degenerate offspring for crime, or to let them starve for their imbecility, society can prevent those who are manifestly unfit from continuing their kind ...Three generations of imbeciles are enough."[123]

The US and European eugenics movements continued to forge ahead. Headlines in the *New York Times* proclaimed eugenicists' belief in the need to limit immigration: "EUGENISTS DREAD TAINTED ALIENS; Believe Immigration Restriction Essential to Prevent Deterioration of Race."[124] The movement's anti-Jewish pronouncements became increasingly vitriolic throughout the 1930s, as the country's economy weakened, with eugenicists openly praising Nazi racial policy.[125] Government committees sought to distinguish between "high-grade Jews" and "low-grade Jews" as they advocated for the exclusion of Jewish refugees from Nazi Germany who failed to demonstrate "real hereditary capacity, sound in physical stamina and of outstanding personal qualities, honesty, decency, common sense, altruism, patriotism and initiative ...."[126] And, indeed, the diminishing numbers of Jewish immigrants to the U.S. reflected the growth of anti-Jewish sentiment. An estimated 120,000 Jews had immigrated to the USA in 1921. After the passage of the Quota Act, that number fell to only 10,000 in 1924, and by 1933, a mere 2000 Jews were permitted to enter.[127]

Jews, it was alleged, were not only "a threat to Christianity," but "Where the Jews are, there trouble is."[128] Father Charles Coughlin, a Detroit priest, became a nationally recognized figure for his vocal diatribes against Jews.[129] Between 1933 and 1941, more than 100 antisemitic organizations were established, in contrast to only

---

Emma Buck, who had also been confined to the Colony; Emma suffered from syphilis, pneumonia, and various other disorders. Even Carrie's daughter Vivian, at the age of three days old, was alleged by the state agency to be feeble-minded. The eugenicist Harry Laughlin administered the newly developed Stanford-Binet intelligence test to both Emma and Carrie and claimed that their test results placed them in the category of imbeciles. Later findings suggest that no evidence existed to support the claim that either Emma or Carrie had any mental deficiency. For a comprehensive review of the circumstances and the court case, see Paul Lombardo, *Three Generations, No Imbeciles: Eugenics, the Supreme Court, and Buck v. Bell.* (Baltimore, MD: Johns Hopkins University Press, 2008). For a discussion of the eugenic underpinnings of intelligence testing, see generally Allan Chase, *The Legacy of Malthus: The Social Costs of the New Scientific Racism* (New York: Alfred A. Knopf, 1977); Stephen J. Gould, *The Mismeasure of Man*, rev. ed. (New York: W.W. Norton, 1996); Leon J. Kamin, *The Science and Politics of IQ* (Mahwah, NJ: Lawrence Erlbaum Associates, 1974).

[123] *Buck v. Bell,* 274 U.S. 207. The Court's decision paved the way for the involuntary sterilization of more than 8000 individuals in Virginia alone.

[124] *New York Times*, September 25, 1921.

[125] "U.S. Eugenicist Hails Nazi Racial Policy," *New York Times*, August 29, 1935.

[126] "Relaxing Quotas for Exiles Sought," *New York Times*, May 4, 1934.

[127] Rita J. Simon, *In the Golden Land: A Century of Russian and Soviet Jewish Immigration in America*. (Westport, CT: Praeger, 1997), 11.

[128] "Jesuit Sees 'World Jews' As Threat to Christianity," *Daily News (New York),* August 30, 1934.

[129] Sheldon Marcus, *Father Coughlin: The Tumultuous Life of the Priest of the Little Flower* (Boston, MA: Little, Brown & Company, 1973); "Fr. Coughlin Denies Charge of Rabbi Wise," *Daily News (New York),* March 25, 1935.

five such organizations prior to that time.[130] Jews were routinely excluded from country clubs, universities, and summer hotels[131] and found their housing opportunities limited by calls for "Christians only."[132]

Even as Louis' family grappled with economic hardship and family disruptions, and the ensuing emotional turmoil, they began to hear news of Hitler's ascendancy and the increasingly harsh restrictions imposed on Jews.

Nazi Eugenics

Even as the US-based race eugenicists began to lose the support of other scientists in the 1930s,[133] Hitler was amassing both scientific and popular support for his eugenic agenda, drawing its parameters from US eugenic policies.[134] Hitler rejected the characterization of Judaism as a religion, characterizing Jews, instead, as a race apart. In a letter to Adolf Gemlich dated September 16, 1919, Hitler commented on the nature of anti-Semitism, explaining that it represented a "political movement" and that Jewry was a racial, rather than religious, association. He explained further that while emotional anti-Semitism leads to pogroms, "rational anti-semitism must lead to a systematic legal opposition and elimination of the special privileges which Jews hold …."[135] And, although significant distinctions existed between religious antisemitism and the Nazi anti-Semitism grounded on racial theory,[136] the Nazis capitalized on the prevalence of antisemitic religious sentiment in their campaign against the Jews.[137]

---

[130] Stanley High, "The Star-Spangled Fascists," *The Saturday Evening Post* 211 (May 27, 1939): 5–7, 70–73.

[131] Julia McCarthy, "Racial Suit Swells Parliman's Jail Woe," *Daily News (New York)*, July 12, 1934; "Anti-Semitism Gains in U.S., Rome Told," *Daily News (New York)*, November 18, 1938, 24.

[132] McCarthy, "Racial Suit"; "Anti-Semitism Gains in U.S., Rome Told." Advertisements for rent and mortgages not infrequently specified that the opportunity was limited to Christians, e.g., *Daily News (New York)*, February 27, 1932, 11.

[133] Andrew S. Winston, "'Jews Will Not Replace Us': Antisemitism, Interbreeding and Immigration in Historical Context," *American Jewish History* 105, no. 1/2 (Jan./April 2021): 8.

[134] US Holocaust Memorial Museum, "Hitler Comes to Power," *Holocaust Encyclopedia*, https://encyclopedia.ushmm.org/content/en/article/hitler-comes-to-power

[135] "Adolf Hitler: On the Annihilation of the Jews," Jewish Virtual Library, https://www.jewishvirtuallibrary.org/adolf-hitler-on-the-annihilation-of-the-jews-spetember-1919

[136] Michael Phayer, *The Catholic Church and the Holocaust, 1930–1965* (Bloomington, IN: University of Indiana Press, 2000).

[137] Stephen W. Eldridge, "Ideological Incompatibility: The Forced Fusion of Nazism and Protestant Theology and Its Impact on Anti-Semitism in the Third Reich," *International Social Science Review* 81, no. 3/4 (2006): 151–165; Robert Michael, *Holy Hatred: Christianity, Antisemitism, and the Holocaust* (New York: Palgrave Macmillan, 2006); Robert N. Proctor, *Racial Hygiene: Medicine Under the Nazis* (Cambridge, MA: Harvard University Press, 1988); Peter Pulzer, *The Rise of Political Antisemitism in German and Austria*, second ed. (Cambridge, MA: Harvard University Press, 1988).

The Nazi racial policy comprised two components. The first required the eradication of groups thought to be defective and/or impure, including Jews, mentally and physically handicapped individuals, and Roma. Jews were believed, for example, to have bad eyes, flat feet, weak backs, and high rates of mental infirmity, sexual deficiency, homosexuality, and various forms of cancer.[138] The second element consisted of protecting and supporting those considered to be Aryan, that is, encompassed within the *Volksdeutsche*, i.e., those "whose language and culture had German origins" but who did not hold German citizenship.[139] Increasingly restrictive and severe anti-Jewish measures were implemented to further effectuate this goal. Despite, or perhaps because of, the ambiguous parameters that defined exactly who constituted a part of the Volksdeutsche, the Nazis relied upon the concept to justify the seizure of property and its transfer to ostensible members of the Volksdeutsche, the implementation of policies designed to maintain the purity of the Volksdeutsche, and the expansion into and acquisition of territories putatively home to members of the Volksdeutsche, some of whom had no apparent affiliation or affinity for their presumed roots.[140]

The medical profession was not immune to these sentiments and prejudices. German social Darwinists relied on the concept of natural selection to argue against the provision of medical care to those who would not survive but for the intervention of medical professionals, asserting that although such attention benefited the individual, it ultimately endangered the race.[141] The development and implementation of "racial hygiene" would simultaneously facilitate control over human reproduction such that the incidence of negative qualities could be reduced and also provide an ostensibly more humane alternative to natural selection. Although a diversity of perspectives existed within the racial hygiene movement, many of the movement's proponents, often physicians, scientists, and professors, advocated the compulsory sterilization and institutionalization of those considered to be defective and reliance on eugenics to address social problems.

By 1930, the racial hygiene movement and the Nordic/Nazi movement, which asserted the supremacy of "Nordic stock," became increasingly aligned.[142] By 1932, not only had the German medical community accepted the concept of racial hygiene as scientific orthodoxy, but the theory was further disseminated through medical

---

[138] Robert N. Proctor, *The Nazi War on Cancer* (Princeton, NJ: Princeton University Press, 1999).

[139] Doris L. Bergen, "The Nazi Concept of 'Volksdeutsche' and the Exacerbation of Anti-Semitism in Eastern Europe, 1939–45," *Journal of Contemporary History* 29 (1994): 569. See also Valdis O. Lumans, *Himmler's Auxiliaries: The Volksdeutsche Mittelstelle and the German National Minorities in Eastern Europe, 1939–1945* (Chapel Hill, NC: University of North Carolina Press, 1993).

[140] Bergen, "Nazi Concept."

[141] Proctor, *Racial Hygiene*.

[142] Proctor, *Racial Hygiene*, 28.

school curricula.[143] Gerhard Wagner, then leader of the German medical profession, is said to have stated in 1933:

> Knowledge of racial hygiene and genetics has become, by a purely scientific path, the knowledge of an extraordinary number of German doctors. It has influenced to a substantial degree the basic world view of the State, and indeed may even be said to embody the very foundations of the present state [*Staatsraison*].[144]

National Socialism, the party of Nazism, became the "political expression of … biological knowledge,"[145] the premier example of "applied biology."[146] Relying on such "knowledge," Hitler had argued that nature's cruelty justified cruelty such that he had "the right to eliminate millions of an inferior race that multiplies like vermin?"[147]

Such ostensibly "objective" science was used to defend existing prejudices, construct a worldview, and implement a legal system that would further perpetuate and reinforce those prejudices. The practice of medicine under the Nazis was shifted from a focus on the individual patient to a focus on the care of the nation; racial hygiene was integral to this new approach, providing a foundation for the distinction between those lives that were considered valuable and those considered "not worth living."[148] Dr. Arthur Guett, a high-ranking health official under the Nazis, declared,

> [T]he ill-conceived "love of thy neighbor" has to disappear …. It is the supreme duty of the … state to grant life and livelihood only to the healthy and hereditarily sound portion of the population in order to secure … a hereditarily sound and racially pure Volk for all eternity.[149]

This reformulated perspective of health care and medicine was reflected in the promulgation of legislation in 1933 to provide for compulsory sterilization on the basis of "eugenic indications" such as mental illness and alcoholism and the corollary establishment of genetic health courts; the implementation of a euthanasia program to reduce the economic burden to the country of institutionalization, contrary

---

143 Florian Bruns and Tessa Chelouche, "Lectures on Inhumanity: Teaching Medical Ethics in German Medical Schools Under Nazism," *Annals of Internal Medicine* 166 (2017): 591–595; Michael H. Kater, *Doctors Under Hitler* (Chapel Hill, NC: University of North Carolina Press, 1989).

144 Proctor, *Racial Hygiene*, 45.

145 Proctor, *Racial Hygiene*, 28.

146 Robert N. Proctor, "Nazi Doctors, Racial Medicine, and Human Experimentation," in *The Nazi Doctors and the Nuremberg Code: Human Rights in Human Experimentation*, edited by George J. Annas and Michael A. Grodin (New York: Oxford University Press, 1992), 19; Robert N. Proctor, "Nazi Science and Nazi Medical Ethics: Some Myths and Misconceptions," *Perspectives in Biology and Medicine* 43, no. 3 (2000): 341.

147 Joachim Fest, *Hitler* (New York: Vintage Books, 1974), 679–680.

148 Proctor, *Racial Hygiene*, 73; Proctor, "Nazi Doctors, Racial Medicine," 24.

149 Robert J. Lifton, *The Nazi Doctors: Medical Killing and the Psychology of Genocide* (New York: Basic Books, 1986), 32.

to German law but authorized by German officials; and the characterization of Jews as a disease within the population, curable only through excision.[150]

Not only medicine, but the physician as well, was to be transformed. The physician was to become a "cultivator of the genes" and "caretaker of the race," "not only … a Party member on the outside, but rather … convinced in their heart of hearts of the biological laws that form the center of his life."[151]

It has been asserted that German physicians were never ordered to participate in these programs, but, empowered to do so, they readily responded affirmatively. Fritz Lenz, a professor of racial hygiene, remarked in 1933:

> Whatever resistance the idea of racial hygiene may have encountered in previous times among German doctors, this resistance exists no longer. The German core [*Kern*] within the medical community has recognized the demands of German racial hygiene as its own; the medical profession has become the leading force in making these demands.[152]

"Difficult moral issues are often ones where social cues are confused, or worse, are pointed in the wrong direction."[153] It is clear that a large proportion of German physicians during this time either felt no moral dilemma existed that required resolution or chose to resolve whatever dilemma they perceived by conforming to what appeared to be the predominant social cues—to adhere to and assist with the practical implementation of Nazi ideology. Their choice fortuitously coincided with their own economic and professional self-interests, at least in the short term.

One might think that Louis and Annie, now living in the Bronx in New York City with their small family, would not be touched by these events occurring far across the Atlantic Ocean. Nothing could be further from the truth.

News from Europe likely reached Louis and Annie not only through the newspapers and radio programs, but also through letters from family members who had remained in Russia and Poland. Then, on November 9, 1938, the Nazis orchestrated a night of terror across Germany, the annexed areas of Austria, and German-occupied areas of Czechoslovakia, looting and destroying Jewish-owned businesses, burning synagogues, forcibly removing Jews from their homes to be beaten and killed—the "night of the broken glass"—*Kristallnacht*, ostensibly in response to the assassination in Paris of the German official Ernst vom Rath. More than 30,000 Jewish males were rounded up and taken to concentration camps. The Jewish community was ordered to pay one billion Reichsmarks as an atonement tax.[154] Ten days later, Father Coughlin broadcast to his millions of US listeners his justification for the Nazi slaughter, charging the Jews with having financed the Russian Revolution,

---

[150] Proctor, *Racial Hygiene*.

[151] Lifton, *Nazi Doctors*, 30, 33, quoting Rudolf Ramm, faculty member, University of Berlin.

[152] Proctor, *Racial Hygiene*, 45.

[153] Peter Mostow, "Like Building on Top of Auschwitz: On the Symbolic Meaning of Using Data from Nazi Experiments and Non-Use as a Form of Memorial," *Journal of Law & Religion* 10 (1993): 403–431.

[154] "Kristallnacht," *Holocaust Encyclopedia*, accessed December 18, 2020. https://encyclopedia.ushmm.org/content/en/article/kristallnacht

causing Germany's economic woes, and controlling the radio and press.[155] Not long after, on February 20, 1939, 19,000 people gathered in Madison Square Garden in the center of New York City, waving banners that proclaimed "Stop the Jewish Domination of America!" and "Wake Up America! Smash Jewish Communism." The gathering, ostensibly organized to celebrate George Washington's birthday, was attended by supporters of Father Coughlin and members of the Christian Front, a far-right anti-Semitic anticommunist political association active in the US during the late 1930s and early 1940s and dedicated to the return of America to its Christian roots through Christian unity.[156]

What might Louis and Annie and their two sons, my father Henry and his brother Irving, both now 20 years old or more, have thought, have felt, reading about this massive gathering, this riotous effort to tarnish Jews, to blame Jews, in essence, to depersonalize and dismiss Jews? Louis, who in the Pale of Settlement had heard of, maybe even witnessed, pogroms against Jews. Perhaps the news of the Madison Square Garden rally brought back memories of these earlier events, perhaps triggering anxiety or fear, perhaps a sense that there would never be any real safety, any real sense of belonging. Perhaps this event compounded the many past incidences of discrimination, of being othered, to be compounded by still more future events, to instill an everlasting distrust of government, of difference, and an inability to see the world as anything but black and white, good and bad, absent of any grey or nuance.

The Jews of Druya could not have anticipated the soon-to-be destruction of their world and, indeed, of life as they knew it. In September 1939, the Nazis invaded Druya, decimating the town's Jewish population.[157]

References

Adams, Mark B. "Eugenics in the History of Science," in *The Wellborn Science: Eugenics in Germany, France, Brazil, and Russia*, edited by Mark B. Adams. New York: Oxford University Press, 1990.

---

[155] Ronald Modras, "Father Coughlin and the Jews: A Broadcast Remembered," *America* (March 11, 1989): 219–222.

[156] Gene Fein, "For Christ and Country: The Anti-Semitic Anticommunism of Christian Front Street Meetings in New York City," *U.S. Catholic Historian* 22, no. 4 (Fall 2004): 37–56; "Socialists Oppose Rally," *Daily News (New York)*, February 20, 1939; "Three Bombs Reported Set for Nazi Rally; 1327 Cops Called," *Daily News (New York)*, February 20, 1939; "Police Break Up Anti-Nazi Crowd at Bund Meeting," *Daily News (New York)*, February 21, 1939; "Police Quell Riots As Foes Besiege Huge Nazi Rally," *Daily News (New York)*, February 21, 1939; "Laugh By Columnist Furnishes Diversion," *Daily News (New York)*, February 21, 1939.

[157] Leonid Smilovitsky, "A Demographic Profile of the Jews in Belorussia from the Pre-War Time to the Post-War Time," *Journal of Genocide Research* 5, no. 1 (2003): 119; Leonid Smilovitsky, *Holocaust in Belorussia, 1941–1944*, translated from *Katastrofia Evreev v Belorusii, 1941–1944* (Tel Aviv, 2000), 127, accessed December 18, 2020. https://www.jewishgen.org/Yizkor/belarus/Belarus.html#TOC

Adams, Mark B. "Eugenics in Russia, 1900–1940." In *The Wellborn Science: Eugenics in Germany, France, Brazil, and Russia*, edited by Mark B. Adams, 153–216. New York: Oxford University Press, 1990.

"Adolf Hitler: On the Annihilation of the Jews," Jewish Virtual Library, https://www.jewishvirtuallibrary.org/adolf-hitler-on-the-annihilation-of-the-jews-spetember-1919.

Allen, Garland E. "The Social and Economic Origins of Genetic Determinism: A Case History of the American Eugenics Movement, 1900–1940 and Its Lessons for Today," *Genetica* 99 (1997): 77–88.

Alroey, Gur. "Out of the Shetl. In the Footsteps of Eastern European Jewish Emigrants to America, 1900–1914," *Leidschrift: Organisatie En Regulering Van Migratie in Di NieuweTijd*22, no. 1 (April 2007): 91–122.

Ansky, S. *The Enemy at His Pleasure: A Journey Through the Jewish Pale of Settlement During World War I*. Translated and edited by Joachim Neugroschel. New York: Henry Holt and Company, 2002.

"Anti-Semitism Gains in U.S., Rome Told," *Daily News (New York)*, November 18, 1938, 24.

"Ashland Branch 'Y' to Hold Baby Contest in Connection with Fall Fair," *The New York Age*, October 18, 1924.

"Ashland Place Branch 'Y' to Hear Miss Eva Bowles; Babies Who Won Awards," *The New York Age*, November 8, 1924.

"Asks Jews to Meet All Attacks Fairly," New York Times (June 19, 1922), 9.

Avrutin, Eugene M. "A Legible People: Identification Politics and Jewish Accommodation in Tsarist Russia," Ph.D. dissertation, University of Michigan, 2004.

Avrutin, Eugene M. "Racial Categories and the Politics of (Jewish) Difference in Late Imperial Russia," *Kritikai: Explorations in Russian and Euroasian History* 8, no. 1 (Winter 2007): 13–40.

Avrutin, Eugene M. and Bemporad, Elissa, Eds. *Pogroms: A Documentary History* (New York: Oxford University Press, 2021.

Bardziński, Filip. "The Concept of the 'New Soviet Man' as a Eugenic Project: Eugenics in Soviet Russia After World War II," *Ethics in Progress* 4, no. 1 (2013): https://doi.org/10.14746/eip-2013-15.

Bergen, Doris L. "The Nazi Concept of 'Volksdeutsche' and the Exacerbation of Anti-Semitism in Eastern Europe, 1939–45," *Journal of Contemporary History* 29, no. 4 (1994): 569–582.

Bernheimer, Charles Seligman. *The Russian Jews in the United States: Studies of Social Conditions in New York, Philadelphia and Chicago, with a Description of Rural Settlements*. Andesite Press, 1905.

"Better Baby Contest," *Daily News*, May 12, 1949.

Bodkin, Karen. *How Jews Became White Folks and What That Says about Race in America*. New Brunswick, NJ: Rutgers University Press, 1990.

Bombay, Amy, Matheson, Kimberly and Anisman, Hymie. "The Intergenerational Effects of Indian Residential Schools: Implications for the Concept of Historical Trauma," *Transcultural Psychiatry* 51, no. 3 (2014): 320–338.

Brothers, Doris. "Traumatic Attachments: Intergenerational Trauma, Dissociation, and the Analytic Relationship," *International Journal of Psychoanalytic Self Psychology* 9, no. 1 (2014): 3–15.

Brumberg, Stephen. *Going to America, Going to School: The Jewish Immigrant Public School Encounter in Turn-of-the-Century New York City*. New York: Praeger, 1986.

Bruns, Florian and Chelouchde, Tessa, "Lectures on Inhumanity: Teaching Medical Ethics in German Medical Schools Under Nazism," *Annals of Internal Medicine* 166 (2017): 591–595.

Buck v. Bell, 274 U.S. 200 (1927).

Budnitskii, Oleg. *Russian Jews between the Reds and the Whites, 1917–1920*, translated by Timothy J. Portice. Philadelphia, PA: University of Pennsylvania Press, 2012.

Campbell, C.G. "The German Racial Policy," *Eugenical News* XXI, no. 2 (March–April 1936): 25–26.

Center for Parent Information and Resources, "Definition." Accessed March 13, 2022, https://www.parentcenterhub.org/downsyndrome/.

Centers for Disease Control and Prevention, "Epidemic Typhus." Accessed March 7, 2022, https://www.cdc.gov/typhus/epidemic/index.html.

Centers for Disease Control and Prevention, "Facts about Down Syndrome," April 6, 2021. Accessed March 13, 2022, https://www.cdc.gov/ncbddd/birthdefects/downsyndrome.html.

"Challenges Ford to Jewish Inquiry," *New York Times*, November 12, 1920, 5.

Chase, Allan. *The Legacy of Malthus: The Social Costs of the New Scientific Racism.* New York: Alfred A. Knopf, 1977.

Cohen, Naomi W. "Antisemitism in the Gilded Age: The Jewish View," *Jewish Social Studies* 41, no. 3/4 (Summer-Autumn 1979): 187–188.

Commons, John R. *Races and Immigrants in America.* New York: Chautauqua Press, 1907.

Connolly, Angela. "Healing the Wounds of Our Fathers: Intergenerational Trauma, Memory, Symbolization and Narrative," *Journal of Analytical Psychology* 56 (2011): 607–626.

Danieli, Yael. "Assessing Trauma Across Cultures From a Multigenerational Perspective." In *Cross-Cultural Assessment of Psychological Trauma and PTSD.* Edited by John P. Wilson and Catherin So-kum Tang, 65–89. Boston, MA: Springer, 2007.

Darwin, Charles. *On the Origin of the Species by Means of Natural Selection, or the Preservation of Favoured Races in the Struggle for Life* (1859). Available at https://www.vliz.be/docs/Zeecijfers/Origin_of_Species.pdf.

Davenport, Charles Benedict. *Eugenics. The Science of Human Improvement by Better Breeding.* New York: Henry Holt and Company, 1910.

De Courcy Ward, Robert. "Race Genetics Problems," *American Breeders' Magazine* 2, no. 3 (Third Quarter 1911): 231.

Dearborn Independent, *The International Jew: The World's Foremost Problem: Being a Reprint of a Series of Articles Appearing in the Dearborn Independent from May 22 to October 2, 1920.* Forgotten Books.

Dinnerstein, Leonard. *Antisemitism in America.* New York: Oxford University Press, Inc., 1994.

"Dr. Rabinow Deplores Bias Against Jew in America, *Jewish Daily Bulletin*, July 6, 1930, 2.

Eldridge, Stephen W. "Ideological Incompatibility: The Forced Fusion of Nazism and Protestant Theology and Its Impact on Anti-Semitism in the Third Reich," *International Social Science Review* 81, no. 3/4 (2006): 151–165.

Emergency Immigration Act of 1921, 42 Stat. 5 (May 19, 1921).

"EUGENISTS DREAD TAINTED ALIENS; Believe Immigration Restriction Essential to Prevent Deterioration of Race," *New York Times*, September 25, 1921.

Farnam, Henry W. "The State and the Poor," *Political Science Quarterly* 3, no. 2 (June 1888): 282–310.

Fein, Gene "For Christ and Country: The Anti-Semitic Anticommunism of Christian Front Street Meetings in New York City," *U.S. Catholic Historian* 22, no. 4 (Fall 2004): 37–56.

Fest, Joachim. *Hitler*. New York: Vintage Books, 1974.

Figes, Orlando. *A People's Tragedy: The Russian Revolution, 1891–1924.* New York: Penguin Books, 1998.

Find a Grave, accessed December 17, 2020, https://findagrave.com/memorial/150906944/louis-fogelman.

Florinsky, Vasily M. Усовершенствование и вырождениечеловечского родаСПБ.: Журнал Дело, 206.

"Fr. Coughlin Denies Charge of Rabbi Wise," *Daily News (New York),* March 25, 1935.

Francis, Janae. "Farmers Lobby for Immigration Reform to Address Labor Shortages," *Christian Science Monitor*, February 23, 2018.

Frazer, Elizabeth F. "Our Foreign Cities: New York," *The Saturday Evening Post* 195, no. 1 (June 16, 1923): 6–7.

Frieden, Nancy M. "The Russian Cholera Epidemic, 1892–1893 and Medical Professionalization," *Journal of Social History* 10, no. 4 (Summer 1977): 538–539.

Frieden, Nancy Mandelker. *Russian Physicians in the Era of Reform*, 1856–1905. Princeton, NJ: Princeton University Press, 1981.

Fudge, Erica. "At the Heart of the Home: An Animal Reading of Mikhail Bulgakov's The Heart of a Dog," *Humanimalia: A Journal of Human/Animal Interface Studies* 1, no. 1 (2009). Available at https://strathprints.strath.ac.uk/29516/1/Humanimalia_Bulgakov.pdf.

Galton, Francis. *Memories of My Life*. London: Methuen, 1898.

Goddard, Henry H. "Mental Tests and the Immigrant," *Journal of Delinquency* II, no. 5 (September 1917): 243–277.

Goldberg, Jacob A. *Incidence of Insanity among Jews*. New York: National Committee for Mental Hygiene, 1922.

Gordy, Miles H. "The German Language in Our Schools," *Educational Review*, October (1918): 257–263.

Gould, Stephen J. *The Mismeasure of Man*, rev. ed. New York: W.W. Norton, 1996.

Graham, Loren. *Lysenko's Ghost: Epigenetics and Russia*. Cambridge, MA: Harvard University Press, 2016.

Graham, Loren R. "Science and Values: The Eugenics Movement in Germany and Russia in the 1920s," *American Historical Review* 82, no. 5 (December 1977): 1133–1164.

Grant, Madison. *The Passing of the Great Race: The Racial Basis of European History*. New York: Charles Scribner's Sons, 1918.

Green, David. "Veins of Resemblance: Photography and Eugenics." *Oxford Art Journal* 7, no. 2 (1984): 3–16.

Guarnaccia, Peter J. "Ataques de Nervios in Puerto Rico: Culture-Bound Syndrome or Popular Illness?" *Medical Anthropology* 15 (1993): 157–270.

Guarnaccia, Peter J., Lewis-Fernandez, Roberto and Rivera Marano, Melissa. "Toward a Puerto Rican Popular Nosology: Nervios and Ataques de Nervios." *Culture, Medicine, and Psychiatry* 27 (2003): 339–366.

Hall, Prescott F. *Immigration and Its Effects upon the United States*, 2nd rev. ed. New York: Henry Holt and Company, 1908.

Hartmann, Edward George. *The Movement to Americanize the Immigrant*. New York: Columbia University Press, 1948.

Hendrick, Burton J. *The Jews in America*. Garden City, NY: Doubleday, Page and Co., 1923.

High, Stanley. "The Star-Spangled Fascists," *The Saturday Evening Post* 211 (May 27, 1939): 5–7, 70–73.

Howe, Irving. *World of Our Fathers: The Journey of East European Jews to America and the Life They Found and Made*. New York: New York University Press, 2005.

Howell, Yvonne. "Eugenics, Rejuvenation, and Bulgakov's Journey into the Heart of Darkness," *Slavic Review* 65, no. 3 (2006): 544–562.

Hutchinson, John F. "Politics and Medical Professionalism After 1905." In *Russia's Missing Middle Class: The Professions in Russian History*, edited by Harley D. Balzer, 89–116. Armonk, NY: M.E. Sharpe, 1996.

Iudin, Tikhon. Психхозы у близнецов. Зхурнал невропатологии и пскхиатрии 7, no. 1 (1907): 68–83.

Iudin, T.I. *Eugenics: A Teaching of Improving Natural Qualities in Humans*. Moscow: Sabashnikovy, 1925.

Jacobson, Matthew Frye. *Whiteness of a Different Color: European Immigrants and the Alchemy of Race*. Cambridge, MA: Harvard University Press, 1998.

"Jesuit Sees 'World Jews' As Threat to Christianity," *Daily News (New York)*, August 30, 1934.

"Jews Denounce Proposed Cut in Immigration," *New York Tribune* (February 14, 1921), 5.

Kahan, Arcadius. "Social Aspects of the Plague Epidemics in Eighteenth Century Russia," Economic Development and Cultural Change 27, no. 2 (Jan. 1979): 255–266.

Kamin, Leon J. *The Science and Politics of IQ*. Mahwah, NJ: Lawrence Erlbaum Associates, 1974.

Kater, Michael H. *Doctors Under Hitler*. Chapel Hill, NC: University of North Carolina Press, 1989.

Kazer, James. A Guide to Willowbrook State School Resources at Other Institutions, October 2005, https://library.csi.cuny.edu/archives/willowbrook-guide

Kazin, Alfred. *A Walker in the City*. New York: Harcourt, Brace, and World, 1951.

Kerenz, Peter. *Civil War in South Russia, 1918–1920*. Berkeley, CA: University of California Press, 1970.

Kerenz, Peter. "Pogroms and White Ideology in the Russian Civil War." In *Pogroms: Anti-Jewish Violence in Modern Russian History*, edited by John D. Klier and Shlomo Lambroza. Cambridge, UK: Cambridge University Press,1992.

Khen, Yu. V. "Unknown Pages of Russian Eugenics," *Herald of the Russian Academy of Sciences* 76, no. 4 (2006): 385–391.

Klier, John D., Lambroza, Shlomo, eds. *Pogroms: Anti-Jewish Violence in Modern Russian History*. Cambridge: Cambridge University Press, 1992.

Krementsov, N.L. От «звериной философии» к медицинскойгенетике:евгеника в

России и Советском Союзе, Историко-биологические исследования 6, no. 2 (2014): 27. ["From Animal Philosophy to Medical Genetics: Eugenics in Russia and the Soviet Union," Studies in the History of Biology].

Krementsov, Nikolai. "Eugenics in Russia and the Soviet Union." In *The Oxford Handbook of the History of Eugenics*, edited by Alison Bashford and Philippa Levine, 413–429. New York: Oxford University Press, 2010.

Krementsov, Nikolai. *Stalinist Science*. Princeton, NJ: Princeton University Press, 1996.

Krementsov, Nikolai. "The Strength of a Loosely Defined Movement: Eugenics and Medicine in Imperial Russia," Medical History 59, no. 1 (2015): 6–31.

Krementsov, Nikolai. *With and Without Galton: Vasilii Florinskii and the Fate of Eugenics in Russia*. Cambridge, UK: Open Book Publishers, 2018.

"Kristallnacht," *Holocaust Encyclopedia*, accessed December 18, 2020. https://encyclopedia.ushmm.org/content/en/article/kristallnacht.

Kropotkin, Peter. "The Sterilization of the Unfit," *Mother Earth* 7, no. 10 (December 1912). Accessed March 6, 2022, https://sniadecki.wordpress.com/2016/02/04/kropotkine-eugenism-en/

Krzhivitskii, Liudvik. Психхическиерасы. St. Petersburg: XX vek), 54–73, 212–223.

Larson, Edward J. *Sex, Race, and Science: Eugenics in the Deep South*. Baltimore, MD: Johns HopkinsUniversity Press, 1995.

"Laugh By Columnist Furnishes Diversion," *Daily News (New York)*, February 21, 1939.

Laughlin, Harry H. *Immigration and Conquest: A Report of the Special Committee on Immigration and Naturalization of the Chamber of Commerce of the State of New York*. New York: Special Committee on Immigration and Naturalization of the Chamber of Commerce of the State of New York, 1939.

Lev-Wiesel, Rachel. "Intergenerational Transmission of Trauma Across the Generations: A Preliminary Study," *Quantitative Social Work* 6, no. 1 (2007): 75–94.

Lifton, Robert J. *The Nazi Doctors: Medical Killing and the Psychology of Genocide*. New York: Basic Books, 1986.

Lombardo, Paul. *Three Generations, No Imbeciles: Eugenics, the Supreme Court, and Buck v. Bell*. Baltimore, MD: Johns Hopkins University Press, 2008.

Loue, Sana. *"My Nerves Are Bad": Puerto Rican Women Managing Mental Illness and HIV Risk*. Nashville, TN: Vanderbilt University Press, 2011.

Lumans, Valdis O. *Himmler's Auxiliaries: The Volksdeutsche Mittelstelle and the German National Minorities in Eastern Europe, 1939–1945*. Chapel Hill, NC: University of North Carolina Press, 1993.

MacMaster, Neil. *Racism in Europe, 1870–2000*. New York: Palgrave, 2001.

Maldari, Lindsay. "Dismantling an Art History of Eugenics: Fascist Propaganda and Visual Rhetoric in *La difesadella raza*." Proceedings of the National Conference on Undergraduate Research (NCUR) 2019, Kennesaw State University, Kennesaw, Georgia, April 11–13, 2019. Accessed April 2, 2024, https://libjournals.unca.edu/ncur/wp-content/uploads/2021/02/2898-Maldari-Lindsay-FINAL.pdf

Malinovich, Nadia. "Race and the Construction of Jewish Identity in French and American Jewish Fiction of the 1920s," *Jewish History* 19, no. 1 (2005): 29–48.

Malzberg, Benjamin. "Mental Diseases among Jews in New York State, 1920 to 1952," *YIVO Annual of Jewish Social Science* 10 (1955): 279–199.

Marcus, Sheldon. *Father Coughlin: The Tumultuous Life of the Priest of the Little Flower*. Boston, MA: Little, Brown & Company, 1973.

Margalit, Ohr and Tziraki-Segal, Chariklia. "Circumcision: Man's Obligation and Woman's Praxis." *NASHIM: A Journal of Jewish Women's Studies & Gender Issues* Fall 5767/2006, no. 12 (Fall 5767/2006): 10–38.

Markel, Howard. *Quarantine! East European Jewish Immigrants and the New York City Epidemics of 1892*. Baltimore, MD: Johns Hopkins University Press, 1997.

Maxwell, Anne. *Picture Imperfect: Photography and Eugenics 1870–1940*. East Sussex, UK: Sussex Academic Press, 2008.

May, James V. *Mental Diseases: A Public Health Problem* (Boston, MA: Richard G. Badger, 1922.

McCarthy, Julia. "Racial Suit Swells Parliman's Jail Woe," *Daily News (New York)*, July 12, 1934.

McGeever, Brendan. *Antisemitism and the Russian Revolution*. Cambridge: Cambridge University Press, 2019.

McLaughlin, A.J. "The American's Distrust of the Immigrant," *Popular Science Monthly* 62 (1903): 230–236.

Meir, Natan. "Jews, Ukrainians, and Russians in Kiev: Interethnic Relations in Late Imperial Associational Life," *Slavic Review* 65, no. 3 (2006): 475–501.

Michael, Robert. *Holy Hatred: Christianity, Antisemitism, and the Holocaust*. New York: Palgrave Macmillan, 2006.

"Modern Jewish History: The Pale of Settlement," Virtual Jewish Library, https://www.jewishvirtuallibrary.org/the-pale-of-settlement.

Modras, Ronald. "Father Coughlin and the Jews: A Broadcast Remembered," *America* (March 11, 1989): 219–222.

Morrison, Toni. *Playing in the Dark: Whiteness and the Literary Imagination* (New York: Vintage-Random House, 1993.

Mostow, Peter. "Like Building on Top of Auschwitz: On the Symbolic Meaning of Using Data from Nazi Experiments and Non-Use as a Form of Memorial," *Journal of Law & Religion* 10 (1993): 403–431.

Murav, Harriet. *Identity Theft: The Jew in Imperial Russia and the Case of Avraam Uri Kovner*. Stanford, CA: Stanford University Press, 2001.

Myrdal, Gunnar. *American Dilemma: The Negro Problem and Modern Democracy*. New York: Harper & Brothers, 1944.

"The Nuremberg Race Laws," *Holocaust Encyclopedia*, United States Holocaust Memorial Museum. Accessed March 6, 2022, https://encyclopedia.ushmm.org/content/en/article/the-nuremberg-race-laws.

O'Brien, Gerald V. "Indigestible Food, Conquering Hordes, and Waste Materials: Metaphors of Immigrants and the Early Immigration Restriction Debate in the United States," *Metaphor and Symbol* 18, no. 1 (2003): 35–38.

"On the Abolition of Religious and National Restrictions," "Оботменевероисповедных и национальныхограничений". Постановление 20 марта 1917 г.

"On the Emergence and Revival of Crazy People, According to the Data of Epigenetics," "О выржедении и возрожденииллудеиподанным еигеник," *Physician's Gazette* [Врачебнаиагазета], 7.

Ontario Ministry of Children, Community and Social Services. "The Words We Used, 1850s–1940s," https://www.mcss.gov.on.ca/en/dshistory/language/1850s_1940s.aspx.

Ontario Ministry of Children, Community and Social Services. "The Words We Used, 1940s–1970s," https://www.mcss.gov.on.ca/en/dshistory/language/1940s_1970s.aspx.

Ordover, Nancy. *American Eugenics: Race, Queer Anatomy, and the Science of Nationalism*. Minneapolis, MN: University of Minnesota Press, 2003.

Parker, A. Warner. "A. Warner Parker on Immigration," *Jewish Immigration Bulletin* (April 1920).

Patten, Simon N. *The Premises of Political Economy: Being a Re-Examination of Certain Fundamental Principles of Economic Science*. New York: Augustus M. Kelley, 1885.

Patterson, K. David. "Typhus and Its Control in Russia, 1870–1940," *Medical History* 37 (1993): 361–381.

Pavlenko, Aneta. "'We Have Room for But One Language Here': Language and National Identity in the US at the Turn of the 20th Century," *Multilingua* 21 (2002): 163–196.

Perlman, Joel. Literacy among the Jews of Russia in 1897: A Reanalysis of Census Data. [Working Paper 1982]. Jerome Levy Economics Institute, 17, available at https://www.levyinstitute.org/pubs/wp182.pdf.

Pernick, Martin S. "Eugenics and Public Health in American History," *American Journal of Public Health* 87, no. 11 (1997): 1767–1772.

Pernick, Martin S. "Taking the Better Baby Contests Seriously," *American Journal of Public Health* 92, no. 5 (May 2002): 707–708.

Petrovsky-Shtern, Yohannon. *Jews in the Russian Army, 1827–1917*. Cambridge, MA: Cambridge University Press, 1988.

Phayer, Michael. *The Catholic Church and the Holocaust, 1930–1965*. Bloomington, IN: University of Indiana Press, 2000.

Pintner, Rudolph. *Intelligence Testing: Methods and Results* (New York: Henry Holt, 1923

Pipes, Richard. *Russia Under the Bolshevik Regime*. New York: Alfred A. Knopf, Inc., 1994.

"Police Break Up Anti-Nazi Crowd at Bund Meeting," *Daily News (New York)*, February 21, 1939.

"Police Quell Riots As Foes Besiege Huge Nazi Rally," *Daily News (New York)*, February 21, 1939.

Polonsky, Antony. *The Jews in Poland and Russia, vol. III, 1914 to 2008*. Oxford, UK: Littman Library of Jewish Civilization, 2012.

Preobrazhenskil, Sergei. Кхирургическаиапрофилактикавырождениа. Современнаииапсикхиатриа 2 (1912): 110–117. [Khirurgicheskaiaprofilaktikavyrozhdeniia; Surgical prevention of degeneration. Sovremennaiiapsikhiatriia (*Modern Psychiatry*) 2 (1912): 110–117]

President's Committee on Mental Retardation, *Mental Retardation: Past and Present* [MR 76]. Washington, D.C.: Author, Jan. 1977.

"Prevention and Control of Epidemics in Russia," УпревлениеделамиПрезидентаРоссийскойФедерации [Boris Yeltson Presidential Library]. Available at https://www.prlib.ru/en/section/1302536.

Problems in Eugenics. London: Eugenics Education Society, 1913.

Proctor, Robert N. "Nazi Doctors, Racial Medicine, and Human Experimentation." In *The Nazi Doctors and the Nuremberg Code: Human Rights in Human Experimentation*, edited by George J. Annas and Michael A. Grodin, 17–31. New York: Oxford University Press, 1992.

Proctor, Robert N. "Nazi Science and Nazi Medical Ethics: Some Myths and Misconceptions," *Perspectives in Biology and Medicine* 43, no. 3 (2000): 335–346.

Proctor, Robert N. *The Nazi War on Cancer*. Princeton, NJ: Princeton University Press, 1999.

Proctor, Robert N. *Racial Hygiene: Medicine Under the Nazis*. Cambridge, MA: Harvard University Press, 1988.

Pulzer, Peter. *The Rise of Political Antisemitism in Germany and Austria*, 2nd ed. Cambridge, MA: Harvard University Press, 1988.

Rapoport, Joe. *Echoes of Druya*. Chestnut Ridge, NY: Town House Press, 1989. https://web.archive.org/web/20041130033655/ http://humanities.edu.ru/db/msg/41430.

Reilly, Philip R. *The Surgical Solution: A History of Involuntary Sterilization in the United States*. Baltimore, MD: Johns Hopkins University Press, 1991.

"Relaxing Quotas for Exiles Sought," *New York Times*, May 4, 1934.

Riis, Jacob. *How the Other Half Lives: Studies among the Tenements of New York*. New York: Penguin Books, 1997 [1890]).

Rogin, Michael. *Blackface, White Noise: Jewish Immigrants in the Hollywood Melting Pot*. Berkeley, CA: University of California Press, 1996.

Rosen, Ruth. *The World Split Open: How the Modern Women's Movement Changed America.* New York, Penguin Books, 2000.
Rosenberg, Alfred. *The Myth of the Twentieth Century*. CreateSpace Independent Publishing Platform, 2011.
Rosenthal, Herman, Lipman, J.G., Rosenthal, Vasili, Wygodsky, L., Mysh, M., Galante, Abraham. "Russia," in *The Jewish Encyclopedia*, vol. 10. New York: Philipson-Samoscz, 1905, 531, Public Domain. Accessed March 23, 2024, https://upload.wikimedia.org/wikipedia/commons/c/c0/Map_showing_the_percentage_of_Jews_in_the_Pale_of_Settlement_and_Congress_Poland%2C_The_Jewish_Encyclopedia_%281905%29.jpg
Ross, Edward A. "The Hebrews of Eastern Europe in America," *The Century Magazine* 88 (Sept. 1914): 785–792
Rothenberg, Paula S. *Race, Class, and Gender in the United States: A Integrated Study*, 7th ed. New York: Worth Publishers, 2007.
Seager, Henry Rogers. "The Minimum Wage as Part of a Program for Social Reform," *Annals of the American Academy of Political and Social Science* 48 (July 1913): 3–12.
Selden, Steven. "'Transforming Better Babies into Fitter Families' Archival Resources and the History of the American Eugenics Movement, 1908–1930." *Proceedings of the American Philosophical Society* 149, no. 2 (June 2005): 199–225.
"Senators Find No Emergency for Alien Ban." *New York Tribune*, January 12, 1921, 9.
"Shalom and B'nai Brith Representatives Score Linking of Immigrants with Crime," *Jewish Daily Bulletin*, September 6, 1932, 3.
"Ship Manifest, SS Latvia, Sailing from Libau, October 18, 1921, arriving at New York City, November 2, 1921." Accessed https://heritage.statueofliberty.org/
Sibley, John. "Kennedy Charges Neglect in State Care of Retarded," *New York Times*, September 10, 1965.
Simon, Rita J. *In the Golden Land: A Century of Russian and Soviet Jewish Immigration in America*. Westport, CT: Praeger, 1997.
Sirotkina, Irina. *Diagnosing Literary Genius*. Baltimore, MD: Johns Hopkins University Press, 2001.
Smele, Jonathan D. *The "Russian" Civil Wars, 1916–1926*. New York: Oxford University Press, 2017.
Smilovitsky, Leonid. "A Demographic Profile of the Jews in Belorussia from the Pre-War Time to the Post-War Time," *Journal of Genocide Research* 5, no. 1 (2003): 117–129.
Smilovitsky, Leonid. *Holocaust in Belorussia, 1941–1944*, translated from *Katastrofia Evreev v Belorusii, 1941 1944* (Tel Aviv, 2000), 127. Accessed December 18, 2020. https://www.jewishgen.org/Yizkor/belarus/Belarus.html#TOC.
"Socialists Oppose Rally," *Daily News (New York)*, February 20, 1939.
Stanchevici, Dmitri. *Stalinist Genetics: The Constitutional Rhetoric of T.D. Lysenko*. Amityville, NY: Baywood Publishing Company, Inc., 2012.
Stanciu, Cristina. "Strangers in America: Yiddish Poetry at the Turn of the Twentieth Century and the Demands of Americanization," *College English* 76, no. 1 (September 2013): 59–83.
Stanislavski, Michael. *Tsar Nicholas I and the Jews: The Transformation of Jewish Society in Russia 1825–1855*. Philadelphia, PA: Jewish Publication Society of America, 1982.
Stoddard, T. Lothrop. "The Pedigree of Judah," *Forum* (March 1926): 326.
"Story on How the 'Protocols' Were Spread Here and Abroad," *New York Times*, September 4, 1921, 81.
"Three Bombs Reported Set for Nazi Rally; 1,327 Cops Called," *Daily News (New York)*, February 20, 1939.
"To Discuss Jew in College," *New York Times*, December 23, 1922, 15.
Turiel, Eliot. "Eugenics, Prejudice, and Psychological Research," *Human Development* (2020), https://doi.org/10.1159/000512492.
United States Holocaust Memorial Museum, "Hitler Comes to Power," *Holocaust Encyclopedia*, https://encyclopedia.ushmm.org/content/en/article/hitler-comes-to-power.

United States House Committee on Immigration and Naturalization, *Biological Aspects of Immigration: Hearings*, 16–17 April 1920.
Ury, Scott. *Barricades and Banners: The Revolution of 1905 and the Transformation of Warsaw Jewry*. Stanford, CA: Stanford University Press, 2012.
"U.S. Eugenicist Hails Nazi Racial Policy," *New York Times*, August 29, 1935.
Van Vechton, C.C. "The Criminality of the Foreign Born," *Journal of Criminal Law and Criminology* 32, no. 2 (1941–1942): 139–147.
Vucinich, Alexander. *Darwin in Russian Thought*. Berkeley, CA: University of California Press, 1988.
Wade, Rex A. *The Russian Revolution, 1917*, 3rd ed. Cambridge: Cambridge University Press, 2017.
Walker, Francis Amasa. *Discussions in Economics and Statistics*, vol. 2. Edited by Davis Dewey. New York: Henry Holt and Co., 1899.
Walker, Francis Amasa. "Restriction of Immigration," *Atlantic* 77 (1896): 822–829.
"Was the Doctor Right?," *Independent*, January 3, 1916, 3.
Weinerman, Eli. "Racism, Racial Prejudice, and Jews in Late Imperial Russia," *Ethnic and Racial Studies* 17, no. 3 (1994): 442–495.
Wilson, J.G. "A Study in Jewish Psychopathology," *Popular Science Monthly* 82 (1913): 264–271.
Winston, Andrew S. "'Jews Will Not Replace Us': Antisemitism, Interbreeding and Immigration in Historical Context," *American Jewish History* 105, no. 1/2 (Jan./April 2021): 1–24.
Wiseman, Hadas, Metzl, Einat and Barber, Jacques P. "Anger, Guilt, and Intergenerational Communication of Trauma in the Interpersonal Narratives of Second Generation Holocaust Survivors," *American Journal of Orthopsychiatry* 76, no. 2 (2006): 176–184.
YIVO Institute for Jewish Research. "Map of Pogroms in Pale of Settlement, Sites of major pogroms, 1881–1884 and 1903–1906." New York: The YIVO Institute for Jewish Research. By permission, from the Archives of The YIVO Institute for Jewish Research.
"Y.M.C.A. Bars Vote to Jew," *New York Herald*, February 20, 1921, 16.
Yumbaco, Brenda J. "State Programs for Mentally Retarded: II. Regional Centers and the Future." *The Catholic Transcript* (October 15, 1964): 13.
Zipperstein, Steven J. *Pogrom: Kishinev and the Tilt of History*. New York: W.W. Norton & Company, Inc., 2018.
Zola, Gary Phillip and Dollinger, Marc eds., *American Jewish History: A Primary Source Reader*. Waltham, MA: Brandeis University Press, 2014.
Zygar, Mikhail. *The Empire Must Die: Russia's Revolutionary Collapse, 1900–1917*. New York: Hachette Book Group, 2017.

Interviews and Personal Communications

Kenneth Fogelman, personal communication, date unknown
Mark Murray, former examiner with the Immigration and Naturalization Service, personal communication, May 12, 2021

Chapter 2
The Intertwined Impact of Economics, Eugenic Policy, and Immigration Restrictions

בלוטאיזנישטקיינוואסער," ("Blutiznishtkanevasser")"Blood is not water," it is said in Yiddish, the equivalent of "Blood is thicker than water," as the saying goes in English. And, even though the English equivalent of it was oft-repeated to us as children, we came to understand that words often do not reflect truth.

On a day like any other day, I was sent to isolation in our basement as punishment for some real or imagined transgression, the nature of which I have long since forgotten. A large chest of drawers stood just to the left of the second flight of stairs as I descended into my purgatory. I had never thought to explore its contents. Usually during my confinements to the basement, I had been more interested in peering out the window that sat at the height of the back yard to see whose feet might be passing by, or exploring the many boxes that were stored in the adjoining boiler room.

Those discoveries though, had ceased to hold any further wonder. On this particular day, I rifled through the secrets held in the chest of drawers. I did not understand the nature of all of the papers there, being all of 7 or 8 years old at the time. What I did understand was my mother's U.S. passport; it said that she had been born in Poland. How could this be, I wondered. She had always said that she had been born in New York City. That was what she had written on all of my school registration forms.

Once released from my confinement, I showed my mother the passport and asked if she had really been born in Poland. She snatched it from my hand, angrily snarling that it was none of my business. It would be many years before I would understand how she and her parents—my grandparents—had come to be in the United States.

S. Loue, *From Public Policy to Family Dynamics*, SpringerBriefs in Social Policy, https://doi.org/10.1007/978-3-031-71818-2_2

Together and Apart: Immigration Policy and the Family

Gitla and Rivka arrived in New York on March 29, 1929, on the S. S. Berengaria from Cherbourg, France[1] (Fig. 2.1). Louis did not yet know them, but years later, Rivka, then Ruth, was to become his daughter-in-law.

Gitla and Rivka had left their Polish town of Lowicz, traveling to join Gitla's husband, Rivka's father, Moische. It had been a long 8 years since Gitla and Rivka had seen him and just 9 months short of the date on which Louis was to arrive from

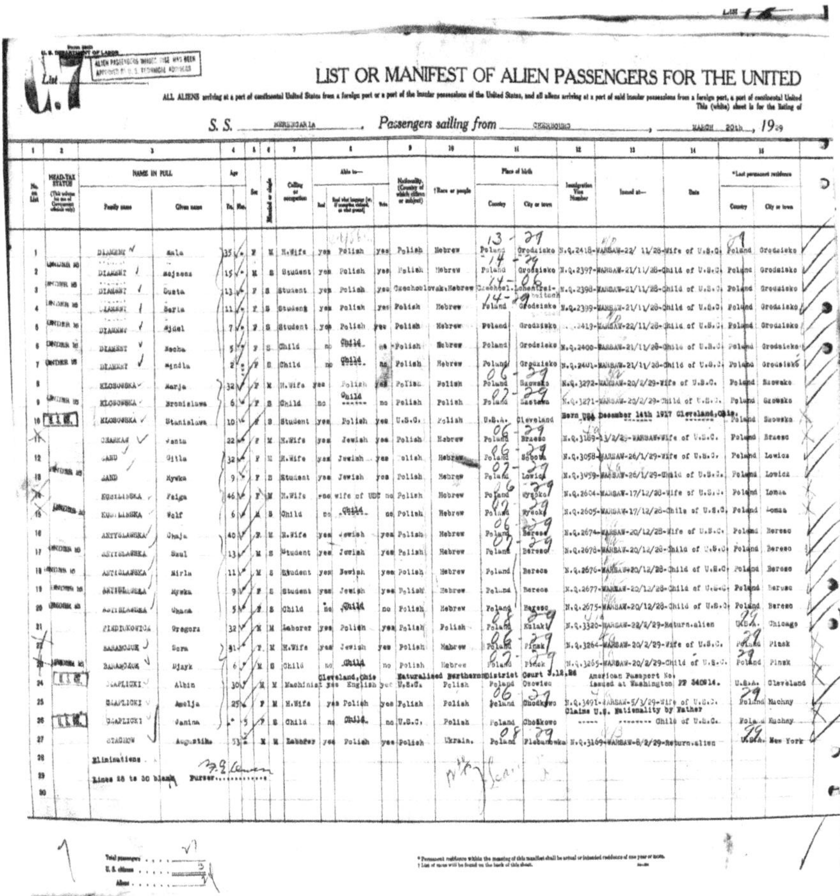

List 7

U. S. DEPARTMENT OF LABOR

ALIEN PASSENGERS WHOSE VISA HAS BEEN APPROVED BY U. S. TECHNICAL ADVISERS

LIST OR MANIFEST OF ALIEN PASSENGERS FOR THE UNITED

ALL ALIENS arriving at a port of continental United States from a foreign port or a port of the insular possessions of the United States, and all aliens arriving at a port of said insular possessions from a foreign port, a port of continental United
This (white) sheet is for the listing of

S. S. BERENGARIA Passengers sailing from CHERBOURG, MARCH 20th, 1929

No. on List	Head-tax status	Family name	Given name	Age	Sex	Married or single	Calling or occupation	Read	Read what language	Write	Nationality	Race or people	Country of birth	City or town of birth	Immigration Visa Number / Issued at / Date	Last permanent residence: Country	City or town
1		DIAMENT	Sala	35	F	M	H.Wife	yes	Polish	yes	Polish	Hebrew	Poland	Grodzisko	N.Q.2418-WARSAW-22/11/28-Wife of U.S.C.	Poland	Grodzisko
2	UNDER 16	DIAMENT	Mojsesz	15	M	S	Student	yes	Polish	yes	Polish	Hebrew	Poland	Grodzisko	N.Q.2397-WARSAW-21/11/28-Child of U.S.C.	Poland	Grodzisko
3	UNDER 16	DIAMENT	Gusta	13	F	S	Student	yes	Polish	yes	Czechoslovak.	Hebrew	Czechosl.	Lohentrei-beitach	N.Q.2398-WARSAW-21/11/28-Child of U.S.C.	Poland	Grodzisko
4	UNDER 16	DIAMENT	Serla	11	F	S	Student	yes	Polish	yes	Polish	Hebrew	Poland	Grodzisko	N.Q.2399-WARSAW-21/11/28-Child of U.S.C.	Poland	Grodzisko
5	UNDER 16	DIAMENT	Ajdel	7	F	S	Student	yes	Polish	yes	Polish	Hebrew	Poland	Grodzisko	N.Q.2419-WARSAW-22/11/28-Child of U.S.C.	Poland	Grodzisko
6	UNDER 16	DIAMENT	Necha	5	F	S	Child	no	Child	no	Polish	Hebrew	Poland	Grodzisko	N.Q.2400-WARSAW-21/11/28-Child of U.S.C.	Poland	Grodzisko
7	UNDER 16	DIAMENT	Mindla	2	F	S	Child	no	Child	no	Polish	Hebrew	Poland	Grodzisko	N.Q.2401-WARSAW-21/11/28-Child of U.S.C.	Poland	Grodzisko
8		KLOSOWSKA	Marja	32	F	M	H.Wife	yes	Polish	yes	Polish	Polish	Poland	Sazowsko	N.Q.3272-WARSAW-20/2/29-Wife of U.S.C.	Poland	Szowsko
9	UNDER 16	KLOSOWSKA	Bronislawa	6	F	S	Child	no	Child	no	Polish	Polish	Poland	Lastawa	N.Q.3271-WARSAW-20/2/29-Child of U.S.C.	Poland	Szowsko
10	U.S.C.	KLOSOWSKA	Stanislawa	10	F	S	Student	yes	Polish	yes	U.S.C.	Polish	U.S.A.	Cleveland	Born USA December 14th 1917 Cleveland, Ohio.	Poland	Szowsko
11		CHARKAS	Janta	22	F	M	H.Wife	yes	Jewish	yes	Polish	Hebrew	Poland	Brzesc	N.Q.3189-13/2/29-WARSAW-Wife of U.S.C.	Poland	Brzesc
12		ZAND	Gitla	32	F	M	H.Wife	yes	Jewish	yes	Polish	Hebrew	Poland	Sobota	N.Q.3058-WARSAW-26/1/29-Wife of U.S.C.	Poland	Lowicz
13	UNDER 16	ZAND	Rywka	9	F	S	Student	yes	Jewish	yes	Polish	Hebrew	Poland	Lowicz	N.Q.3059-WARSAW-26/1/29-Child of U.S.C.	Poland	Lowicz
14		KOSILINSKA	Faiga	46	F	M	H.Wife	yes	wife of USC	no	Polish	Hebrew	Poland	Wysoko	N.Q.2604-WARSAW-17/12/28-Wife of U.S.C.	Poland	Lomza
15	UNDER 16	KOSILINSKA	Wolf	6	M	S	Child	no	Child	no	Polish	Hebrew	Poland	Wysoko	N.Q.2605-WARSAW-17/12/28-Child of U.S.C.	Poland	Lomza
16		ANTYSLAWSKA	Chaja	40	F	M	H.Wife	yes	Jewish	yes	Polish	Hebrew	Poland	Bereso	N.Q.2674-WARSAW-20/12/28-Wife of U.S.C.	Poland	Bereso
17	UNDER 16	ANTYSLAWSKA	Saul	13	M	S	Student	yes	Jewish	yes	Polish	Hebrew	Poland	Bereso	N.Q.2678-WARSAW-20/12/28-Child of U.S.C.	Poland	Bereso
18	UNDER 16	ANTYSLAWSKA	Mirla	11	M	S	Student	yes	Jewish	yes	Polish	Hebrew	Poland	Bereso	N.Q.2676-WARSAW-20/12/28-Child of U.S.C.	Poland	Bereso
19	UNDER 16	ANTYSLAWSKA	Rywka	9	F	S	Student	yes	Jewish	yes	Polish	Hebrew	Poland	Bereso	N.Q.2677-WARSAW-20/12/28-Child of U.S.C.	Poland	Bereso
20	UNDER 16	ANTYSLAWSKA	Chana	5	F	S	Child	no	Child	no	Polish	Hebrew	Poland	Bereso	N.Q.2675-WARSAW-20/12/28-Child of U.S.C.	Poland	Bereso
21		PIEDRUKOWICA	Gregorz	32	M	M	Laborer	yes	Polish	yes	Polish	Polish	Poland	Kulaki	N.Q.3320-WARSAW-22/2/29-Return.alien	U.S.A.	Chicago
22		BARANOJUK	Sora	31	F	M	H.Wife	yes	Jewish	yes	Polish	Hebrew	Poland	Pinsk	N.Q.3264-WARSAW-20/2/29-Wife of U.S.C.	Poland	Pinsk
23	UNDER 16	BARANOJUK	Djayk	6	M	S	Child	no	Child	no	Polish	Hebrew	Poland	Pinsk	N.Q.3265-WARSAW-20/2/29-Child of U.S.C.	Poland	Pinsk
24	U.S.C.	CZAPLICKI	Albin	30	M	M	Machinist	yes	English	yes	U.S.C. Naturalized Northern District Court 3.12.26 Cleveland, Ohio	Polish	Poland	Osowiec	American Passport No. issued at Washington PP 340914.	U.S.A.	Cleveland
25		CZAPLICKI	Amelja	25	F	M	H.Wife	yes	Polish	yes	Polish	Polish	Poland	Chodkowo	N.Q.3491-WARSAW-5/3/29-Wife of U.S.C. Claims U.S. Nationality by Father	Poland	Kuchny
26	U.S.C.	CZAPLICKI	Janina	5	F	S	Child	no	Child	no	U.S.C.	Polish	Poland	Chodkowo	----- -------- Child of U.S.C.	Poland	Kuchny
27		STACHOW	Augustin	53	M	M	Laborer	yes	Polish	yes	Polish	Ukrain.	Poland	Plebanowka	N.Q.3169-WARSAW-8/2/29-Return.alien	U.S.A.	New York
28		Eliminations															
29		Lines 28 to 30 blank	Purser............														
30																	

Total passengers ....
U. S. citizens 3
Aliens ....

* Permanent residence within the meaning of this manifest shall be actual or intended residence of one year or more.
† List of races will be found on the back of this sheet.

Fig. 2.1 Excerpt from Passenger Manifest, S.S. Berengia, Sailing from Cherbourg, France, arriving in New York March 29, 1929. (Source: https://heritage.statueofliberty.org)

[1] The Statue of Liberty—Ellis Island Liberty Foundation, Inc., https://heritage.statueofliberty.org/passenger-details/

Libau with his small family. Contrary to conventional wisdom suggesting that most Jews immigrating to the USA during this period immigrated together with their family members,[2] Moische had immigrated to the USA alone, journeying through Rotterdam, leaving on the S. S. Ryndham on January 25, 1921[3] (Fig. 2.2), arriving only a few months before the Congressional passage of the Emergency Quota Act on March 19, 1921.[4]

The 1921 Act restricted the annual number of individuals that could be admitted to the USA from the Eastern Hemisphere to 3% of the number of individuals from each country that were living in the USA according to the 1910 census. The Act was touted as a temporary measure, designed to afford Congress sufficient time to devise an alternative approach to control immigration. The more "permanent" measure—the 1924 Immigration Act—was passed by Congress on April 18, 1924.[5] That legislation, which maintained the national origins quota system that had been introduced by the 1921 immigration legislation but with the 1890 census as a basis, ultimately led to a drastic reduction in the admission of Jews from eastern European countries. The legislation represented a long-sought victory for eugenicists, nativists, and immigration restrictionists, who had variously characterized Jews as radicals, as unassimilable, and as the cause of the country's economic woes. During the intervening years between Moische's arrival and that of Gitla and Rivka, Moische had become a US citizen and had renamed himself Morris. Gitla and Rivka were able to overcome the Act's restrictions as the immediate family members of a US citizen (see Fig. 2.3, family photo just prior to Moische's immigration to the USA and Fig. 2.4, photo of Moische, Gitla, and Rivka prior to Moische's departure).

At the time that Gitla and Rivka left their home, Lowicz was a thriving urban center in a largely agricultural area, with a large, vibrant Jewish community that comprised one-quarter of the town's inhabitants. The Jewish community could boast of a Jewish library that had been established during the closing years of Russian tsarist rule, a Yiddish weekly, and numerous houses of study and prayer.[6] The sleepy town on the Bzura river bustled with heightened activity on Tuesdays and Fridays, when peasants from the nearby villages arrived at the marketplace[7] (see Fig. 2.5, depicting marketplace).

---

[2] Gur Alroey, "Out of the Shetl," in *Bread to Eat, Clothes to Wear: Letters from Jewish Migrants in the Early Twentieth Century* (Detroit, MI: Wayne State University Press, 2011), 91.

[3] The Statue of Liberty—Ellis Island Liberty Foundation, Inc.

[4] Pub. L. 67–5, 42 Stat 5. The legislation is also known as the Emergency Immigration Act of 1921, the Immigration Restriction Act of 1921, and the Johnson Quota Act.

[5] Pub. L. 68–139, 43 Stat. 153. The Act is also known as the Johnson-Reed Act.

[6] G. Shaiak, "Lowicz: A Short History of the City and Its Jews," in *Lowicz; A Town in Mazovia Yizkor Book (Lowicz, Poland)*, ed. G. Shaiak. trans. from *Lowicz; ir be-Mazovia u-seviva, seferzikaron* (Tel Aviv, 1966), 6. Accessed December 14, 2020. https://www.jewishgen.org/yizkor/lowicz/Lowicz.html#TOCE

[7] Miriam Spicehandler, "Market Day," in *Lowicz; A Town in Mazovia Yizkor Book (Lowicz, Poland)*, ed. G. Shaiak. trans. *Lowicz; ir be-Mazovia u-seviva, seferzikaron* (Tel Aviv, 1966). Accessed December 14, 2020. https://www.jewishgen.org/yizkor/lowicz/Lowicz.html#TOCE

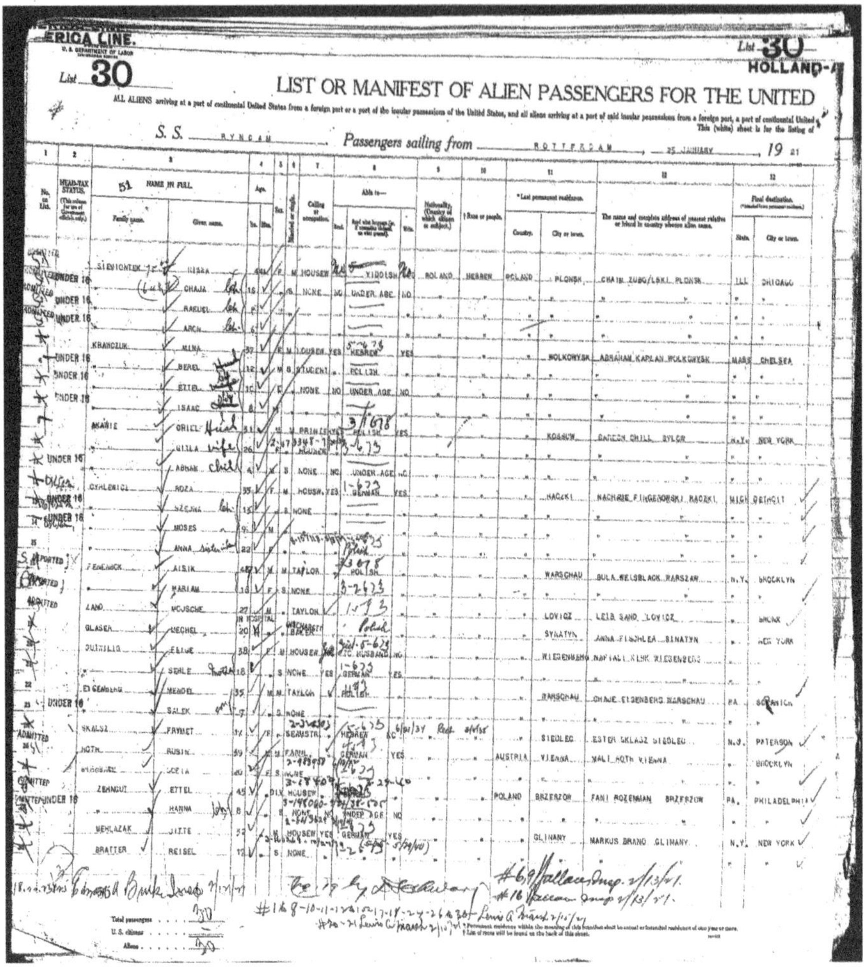
ERICA LINE.

List 30

LIST OR MANIFEST OF ALIEN PASSENGERS FOR THE UNITED

HOLLAND-

S. S. RYNDAM Passengers sailing from ROTTERDAM, 25 JANUARY, 1921

Fig. 2.2 Excerpt from Passenger Manifest, S.S. Ryndham, Sailing from Rotterdam, January 25, 1921. (Source: https://heritage.statueofliberty.org/)

Moische was a tailor, from a long line of tailors, and he had no trouble putting his skills to use shortly after his arrival in New York. He had represented himself as a single male on the ship's manifest,[8] perhaps to increase his chances of passing the immigration inspection that would greet him on his arrival into New York. Immigration policy at the time favored single men with skills, such as Moische had;[9] they would be able to find employment relatively easily in the burgeoning

[8] The Statute of Liberty—Ellis Island Liberty Foundation, Inc.

[9] Bina Kalola, "Immigration Laws and the Immigrant Woman: 1885–1924," *Georgetown Immigration Law Journal* 11, no. 3 (Spring 1997): 553–600.

Fig. 2.3 The Zand Family, circa 1920. From left to right (front row): Moische, Avrumel, Rachel Leah, Fradel, Leib, Model; (back row) Rivka, Gitla, Sarah Rifka. The photo was likely taken shortly before Moische left for the USA. (Source: Author)

garment trade and would not be weighed down themselves or burden the community with a wife and child who needed support. Or perhaps Morris fancied himself a new identity and a new life, one with fewer responsibilities and headaches. After all, he had married Gitla, four years his senior, when he was only 18, and Rivka had been born a short 10 months or so after their wedding.[10] Too, Gitla was not an easy person to live with, with her constant anxieties and worries, her ever-present nagging and hand-wringing.

Once in New York City, Morris did, indeed, build a new and entirely different life for himself, one that was presumably congruent with his values. He became a union organizer in the International Ladies' Garment Workers' Union (ILGWU), even participating in their 1923 strike. Perhaps he became lonely, perhaps he fancied himself a ladies' man—Morris married yet again, to Dora, who had immigrated to the USA from Pułtusk, Poland, on November 8, 1921, coincidentally just a few short weeks before Louis and his family had arrived at Ellis Island. Morris hadn't bothered to divorce Gitla; family murmurings hint at Morris' later conviction for bigamy and a one-year stint in jail, but that would be a few years hence if, in fact, the tale was true.

[10] Extract from record of marriage, Act of Marriage A59/1918, City of Lowicz.

Fig. 2.4 Moische, Gitla, and Rivka in Lowicz, circa 1920. (Source: Author)

Or perhaps Moische—Morris—was experiencing what one would now call an identity crisis. It has been hypothesized that psychosocial growth and development occur in stages, each of which is associated with a psychosocial crisis.[11] In this context, a "crisis" is conceived of as "a turning point for better or worse" to which the individual can respond either adaptively or maladaptively.[12] The extent to which an individual is able to resolve successfully each such crisis depends upon his or her experiences during earlier stages of development. Each stage marks the development of a different facet of the individual's identity in relation to the external social world; the component parts of the individual ultimately give rise to the whole individual (known as epigenetic theory). The successful resolution of the crisis at a particular stage of development results in the development of a basic psychological strength or virtue at that stage.[13]

[11] Erik Erikson, *The Life Cycle Completed (Extended Version)* (New York: W.W. Norton & Company, 1997).

[12] Erik Erikson, *Insight and Responsibility* (New York: W.W. Norton & Company, 1964), 139.

[13] Erikson hypothesized that adult development progresses through 9 stages: infancy, early childhood, play age, school age, adolescence, young adulthood, adulthood, old age, and gerotranscen-

Fig. 2.5 Marketplace in Lowicz. (By permission, from the Archives of The YIVO Institute for Jewish Research, New York)

At the time that Moische had immigrated to the USA, he had reached the age and stage of young adulthood, a period during which individuals must develop the capacity to become intimate with and care about others. The challenge is to be able to commit oneself in a relationship that may require compromise and sacrifice. Perhaps Moische's entry into a new relationship may have signified his emotional growth and a newfound ability to love. How he understood, integrated, and internalized his seeming abandonment or betrayal of Gitla and Rivka is unknown. "Part of [one's] identity must be accounted for in that communality within which an individual finds himself" and there may be "fragments that the individual had to submerge in himself as undesirable or irreconcilable or which his group has taught him to perceive as the mark of fatal "difference" in sex role or race in class or religion."[14]

An individual's identity is anchored in their sense of their corporeal existence and their "subjective self-sameness."[15] Identity is a coherent sense of self. It depends upon the awareness that one's endeavors and one's life make sense, that they are

dence. Erikson, *The Life Cycle*; Erik Erikson, *Childhood and Society* (New York: W.W. Norton & Company, 1951).

[14] Erik Erikson, *Life History and the Historical Moment* (New York: W.W. Norton & Company, 1975), 19–20.

[15] Salman Akhtar, *Immigration and Identity: Turmoil, Treatment, and Transformation* (Northvale, NJ: Jason Aronson Inc., 1999), 63.

meaningful in the context in which life is lived. It depends also upon stable values, and upon the conviction that one's actions and values are harmoniously related.[16]

Whatever Morris' reasons may have been for not sending for them sooner, for having created a new life for himself, Gitla did not see her husband and Rivka her father until 1929, when Morris finally sent for them in response to the incessant pleadings of a New York-based relative who had become aware of Moische's relationship with Dora and who knew Gitla. Rivka idolized her father, who likely was the source of her resilience, rooted "in the sense of being understood by and existing in the mind and heart of a loving, attuned, and self-possessed other."[17] But when Moische left her in Poland, he had left her to fend for herself from the abuses that Gitla rained down on her each day, screaming at her that she would amount to something only when Gitla grew hair on her palms, and the torment inflicted by her grandfather Leib, who used her as pin cushion, sticking pins and needles into her whenever he thought it warranted. Rivka's grandmother had to have been aware of her husband's torture of the little girl. She did nothing to shield her from the abuse, not from her husband and not from Gitla. Perhaps she was too scared for herself, afraid that her husband's wrath would be visited upon her if she were to intervene. Or perhaps she, like Gitla, resented Rivka, not because of anything the little girl was or did, but simply because she was there, an object on which to direct all of her frustrations, all of her unhappiness, all of her disappointments, all of her anxieties (see Fig. 2.6, photo of Rivka in Poland).

Rivka's experiences differed markedly from the oft-repeated portrait of the loving, peaceful, supportive, violence-free Jewish home. Family life was supposed to be "the key to happiness,"[18] and for those raised in a shtetl as Rivka was, with parents who "love you always no matter what happens" and the "quality of affection [may be] so taken for granted," with parental love that is both unconditional and unbreakable.[19] Indeed, this image of the Jewish home serves as the foundation for the stereotype of the "*Yiddishehmammeh*," the Jewish mother, who can never do enough for her children and worries about them incessantly,[20] who "suffocated her

[16] Allen Wheelis, *The Quest for Identity* (New York: W.W. Norton & Co., Inc., 1958), 19. Identity has also been conceived of as being of common origin or experience. Daniel Boyarin and Jonathan Boyarin, "Diaspora: Generation and the Ground of Jewish Identity," *Critical Inquiry* 19 (1993): 693–725. Others have asserted that identity is relational, continuously evolving, and dependent upon difference. Lawrence Grossberg, "Identity and Cultural Studies: Is That All There Is?" *Questions of Cultural Identity*, ed. Stuart Hall and Paul du Gay, 87–107 (London: Sage Publications, 1996).

[17] Diana Fosha, "Dyadic Regulation and Experiential Work with Emotion and Relatedness in Trauma and Disordered Attachment," in *Healing Trauma: Attachment, Trauma, the Brain and the Mind*, ed. Marion F. Solomon and Daniel J. Siegel (New York: Norton, 2003), 228.

[18] Hayim Halevy Donim, *To Be a Jew: A Guide to Jewish Observance in Contemporary Life* (New York: Basic Books, Inc., 1972), 121.

[19] Mark Zborowski, *Life Is with People: The Jewish Little-Town of Eastern Europe* (New York: International University Press, Inc., 1952), 293.

[20] Zborowski, *Life with People*, 294. As one writer notes, "Jews and food have long been linked in the American audiovisual imagination, at least as far back ad Gertrude Berg's Goldbergs radio and

Fig. 2.6 Rivka, Poland, unspecified date 1920s. (Source: Author)

family, but especially her children, with food and nurturance that made giving and receiving a poisoned act; [who instilled] guilt based on the supposed (and yet impossible) demand for total loyalty ….[21] As Howe has noted, "Jewish folklore had elevated the mother to be a figure of sanctioned tenderness, and now, together with a somewhat abrasive capacity for battling whatever might threaten the family, she became an object of sentimental veneration."[22]

Few scholars have examined or acknowledged the discord that lay below the surface of this illusion. Although Zborowski describes what would now often be considered to be child abuse, he concludes, nevertheless, that "All the sacrifice, all

---

TV show (1929–50, 1949–56), which itself built on the archetype of the emotionally and gastronomically nurturing *Yiddische mama*." Nathan Abrams, "Digesting Woody: Food and Foodways in the Movies of Woody Allen," in *Woody on Rye: Jewishness in the Films and Plays of Woody Allen*, edited by Nathan Abrams, Vincent Brooks, and Marat Grinberg, (Waltham, MA: Brandeis University Press, 2014), 215.

[21] Riv-Ellen Prell, *Fighting to Become Americans: Jews, Gender, and the Anxiety of Assimilation*. Boston, MA: Beacon Press, 1999): 144.

[22] Irving Howe, *World of Our Fathers: The Journey of East European Jews to America and the Life They Found and Made* (New York: New York University Press, 2005),176.

the suffering, all the solicitude pile up into a monument to parental love, the dimensions of which define the vastness of filial indebtedness."[23]

> The father is the one who wields the lash, for major offenses are referred to him and he administers punishment to scale .... Mother's slaps and scoldings are of a different order ... [T]hey often reflect the domestic climate rather than one's own offense. They come easily and are easily forgotten, while paternal discipline is less frequent and is long remembered.[24]

Rivka almost never spoke of these experiences, experiences that would now be recognized as abusive, an "adverse childhood experience (ACE),"[25] and maybe even prompt intervention by a social service or government agency, but then were likely to be seen as only a manner of disciplining the child to be obedient and fearful, if not respectful, of authority. On the rare occasions when she spoke of them as an adult, it was usually to compare her experiences of suffering to those of others, as if to say that her life was worse and she had suffered more, or to demonstrate how well she was doing despite her adversities. Now, the verbal and physical abuse that Rivka experienced is considered a form of trauma, one that is linked to toxic or chronic stress and depression in later life, effects that can be passed on to the next generation.[26] The adult Rivka—Ruth—would never be diagnosed with clinical depression—"Psychiatrists are for crazy people. What good are they?!"—but her negativity, her frequent lethargy, her infrequent laughter and smiles with her family, her facile ability to cast blame and disparage others, her readiness to take offense where none seemed to have been intended, all spoke to a profound unhappiness, despite her protestations.

While we do not know what meaning the child Rivka assigned to these abusive experiences, we do know that the meaning given to them plays a critical role in a child's development. Many children believe that they are responsible for having provoked their abuse,[27] a perspective that allows feelings of helplessness to be replaced with feelings of control, forestalling a sense of helplessness.[28] Too, "[t]he child needs to hold on to an image of the parent as good in order to deal with the intensity of fear and rage which is the effect of the tormenting experiences."[29]

---

[23] Zborowski, *Life with People*, 297.

[24] Zborowski, *Life with People*, 296.

[25] Hanna Lee, Kristin S. Slack, Lawrence M. Berger, Rebecca S. Mather, and Rebecca K. Murray, "Childhood Poverty, Adverse Childhood Experiences, and Adult Health Outcomes," *Health & Social Work* 46, no. 3 (August 2021): 159–170.

[26] Centers for Disease Control and Prevention, "Preventing Adverse Childhood Experiences," April 3, 2020, accessed December 14, 2020. https://www.cdc.gov/violenceprevention/aces/fastfact.html

[27] Onnon van der Hart, Ellert R.S. Nijenhuis, and Kathy Steele, *The Haunted Self: Structural Dissociation and the Treatment of Chronic Traumatization* (New York: W.W.W Norton & Company, 2006), 28. See also Anna C. Salter, *Transforming Trauma: A Guide to Understanding and Treating Adult Survivors of Child Sexual Abuse.* (Thousand Oaks, CA: Sage, 1995).

[28] Ann W. Burgess and Lynda L. Holstrom, "Adaptive Strategies in Recovery from Rape." *American Journal of Psychiatry 1*36 (1979): 1278–1282.

[29] Patricia P. Reiker and Elaine Carmen, "The Victim-to-Patient Process: The Disconfirmation and Transformation of Abuse," *American Journal of Orthopsychiatry* 56 (1986): 368.

"[T]he meaning of a threat or traumatic loss can lead to a major shift in an individual's perceptual sensitivities."[30] Perhaps it was these experiences that sowed the seed of shame, feelings of inadequacy, and difference that would grow through the years from repeated harangues, from antisemitism in both her old and her new country, and from the constant readjustments that life demanded of her. We know that shame can be induced by childhood experiences with family and peers, by culture and society.[31] Whether Rivka ever became consciously aware of her feelings of shame is unclear, but many years in the future, her children would recognize that she had felt shame. "The shame that accompanies such personal violations as … abuse is so painful that it is frequently dissociated: Victims may be unaware of its presence, and yet it comes to dominate their interactions with the environment," leading at times to disorganized patterns of engagement.[32] Perhaps it was these early experiences, compounded by later economic difficulties, lost educational opportunities, discriminatory experiences, and subsequent historical events that were responsible in later years for Rivka's general distrust and suspiciousness, her verbal and physical aggression toward her children, and her difficulties connecting to her children emotionally.[33]

Perhaps Gitla, too, had suffered as a child, and reenacted with her daughter the abuse that had been showered on her. Born in 1897,[34] she would have suffered through the famine, hunger, and epidemics that befell Lowicz' Jews during World War I, leading to the opening of soup kitchens to feed the community.[35] Certainly as an adult, she must have suffered. Later, in 1921, Moische had left her with a child of only 1–1/2 years, perhaps because of Lowicz' deteriorating economy,[36] perhaps saying that he would begin a better life for them in America and send for her soon. Might she have felt abandoned, as so many other Jewish women had been by their

[30] Alexander C. McFarlane and Rachel Yehuda, "Resilience, Vulnerability, and the Course of Posttraumatic Reactions," in *Traumatic Stress: The Effects of Overwhelming Experience on Mind, Body, and Society*, ed. Bessel A. van der Kolk, Alexander C. McFarlane, and Lars Weisaeth (New York: Guilford Press, 1996), 164. See also Bessel van der Kolk, "The Compulsion to Repeat the Trauma: Re-enactment, Revictimization and Masochism." *Psychiatric Clinics of North America* 12, no. 2 (1989): 389–411.

[31] Salman Akhtar, workshop "Fear, Greed, Shamelessness Hatred Guilt," Cape Cod Institute, Eastham, MA, August 20–24, 2018; lecture on shame, August 22, 2018.

[32] Bessel A. van der Kolk and Alexander C. McFarlane, "The Black Hole of Trauma," in *Traumatic Stress: The Effects of Overwhelming Experience on Mind, Body, and Society*, ed. Bessel A. van der Kolk, Alexander C. McFarlane, and Lars Weisaeth (New York: Guilford Press, 1996), 15.

[33] For a discussion of the development of affect regulation and dysregulation, see Bessel A. van der Kolk, "The Complexity of Adaptation to Trauma: Self-Regulation, Stimulus Discrimination, and Characterological Development,," in *Traumatic Stress: The Effects of Overwhelming Experience on Mind, Body, and Society*, ed. Bessel A. van der Kolk, Alexander C. McFarlane, and Lars Weisaeth (New York: Guilford Press, 1996), 182–213.

[34] Petition for Naturalization of Gussie Zand.

[35] Martin Zand, *The Zands of Lowicz: A Family History* (Rochester, NY: private printing, 2013), 13.

[36] Zand, *The Zands of Lowicz*, 13.

husbands who had left them for better lives and new identities in America?[37] Or, perhaps she took solace in the knowledge that other women were experiencing the same fate as she, a fate that likely prompted feelings of embarrassment and shame. Family abandonment within the Jewish immigrant community was not an inconsequential problem. Between 1900 and 1922 it was said to have reached crisis proportions; 15% or more of the total annual disbursements of Jewish charitable organizations was allocated to the support of abandoned families.[38] Abandoned wives in the USA, Russia, and Galicia often wrote letters to the New York-based Yiddish newspaper *Forverts* (*Forward*), seeking advice about their situations.[39] The *Forverts* even ran a feature titled "Gallery of Missing Husbands," with photographs of the missing men and their wives' pleas for them to return to their homes.[40]

Numerous theories abounded as to the root causes. Some men, it was thought, left their families after their wives and children arrived in the USA because they discovered that they could no longer relate to them.[41] Some, like Moische, took new wives in the USA, only to be confronted by their former wives upon their arrival in the USA from Eastern Europe.[42] Abandonment was frequently attributed to tensions within the marital household, which was sometimes exacerbated by the presence of boarders in the home, frequently a necessity if the family was to be able to pay the rent.[43]

---

[37] Anna R. Igra, *Wives Without Husbands: Marriage, Desertion, & Welfare in New York, 1900–1935* (Chapel Hill, NC: University of North Carolina Press, 2007), 13–15. Desertion of Jewish wives by their husbands has been an issue at least as far back as the sixteenth century. Chapter 17 of the *Shulchan Arukh*, the Code of Jewish Law, authored by Joseph Karo in 1563, deals specifically with the problem of the abandoned wife, the *agunah*. An agunah is a woman whose husband has disappeared; she is tied to him until he either grants her a divorce according to Jewish law or his death is established. Jewish courts are not empowered to grant a divorce if the husband disappears or refuses. In some cases, the court may choose to relax the laws of evidence to permit the woman to obtain the requested divorce. Isaac Klein, *A Guide to Jewish Religious Practice* (New York: Jewish Theological Seminary of America, 1979), 450–451.

[38] Reena Sigman Friedman, "'Send Me My Husband Who Is in New York City': Husband Desertion in the American Jewish Immigrant Community 1900–1926," *Jewish Social Studies* 44, no. 1 (1982): 5.

[39] For examples of these letters, see Isaac Metzker, ed., *A Bintel Brief: Sixty Years of Letters from the Lower East Side to the Jewish Daily Forward* (New York: Doubleday & Company, Inc., 1971); Isaak Metzker, ed., *A Bintel Brief, Vol. II: Letters to the Jewish Daily Forward, 1950–1980*, trans. Bella S. Metzker and Diana Shalet Levy (New York: Viking Press, 1981); Liana Finck, *A Bintel Brief: Love and Longing in Old New York* (New York: HarperCollins Books, 2014). For a discussion about the Bintel Brief advice column and its author, Abraham Cahan, see Magdalena Ewa Bier, *How to Become Jewish Americans?: The A Bintel Brief Column in Abraham Cahan's Yiddish Forverts* (Frankfurt am Main, Germany: Peter Lang, 2015).

[40] Howe, *World of Our Fathers*, 179.

[41] Gur Alroey, *Bread to Eat*, 73–74.

[42] Friedman, "Husband Desertion," 6.

[43] Friedman, "Husband Desertion," 4.

Some men, finding themselves unable to support their families financially due to the seasonal nature of their employment, left in search of work, never to return.[44] Some observers laid the blame for husbands' wanderings directly at the feet of the women who had been abandoned, suggesting that the women had failed to remain attractive to,[45] create a pleasant home for,[46] or provide adequate sexual attention to their hard-working husbands.[47]

The data utilized in a 1926 study of a sample of 423 "desertion" cases in 1926 from among 1053 suggest that men's abandonment of their families, at least in some cases, may have been one element in their developmental quest for fulfillment and identity. Many of the 343 cases in which the husband's age at marriage was known were adolescents at the time that they were wed; a large number of these marriages had occurred prior to the men's immigration.[48] In the 350 cases in which the husband's age at abandonment was recorded, one-third of the husbands left their wives while still in their teens or twenties, having been married in some cases for only a few years or less, a situation not unlike that between Moische and Gitla. It is also possible that some of these men left in an effort to build a rewarding life in which they could find meaning, rather than pursuing a life trajectory to which they had consented,[49] but which they may not have welcomed or even wanted.

Still other men were perceived as having acceded to the demands or having welcomed the mores of their new environment, embracing new customs and behaviors before their wives, often resulting in marital friction.[50] Men's self-indulgence in this age of rising materialism and consumerism was targeted as one such unwelcome behavior.[51] The lawyer and social worker Bernhard Rabbino, a Jewish immigrant himself, opined with respect to such "deserting" men:

---

[44] Friedman, "Husband Desertion," 6; Alroey, *Bread to Eat*, 73–74.

[45] Joanna C. Colcord, *Broken Homes: A Study of Family Desertion and Its Social Treatment* (New York: Russell Sage Foundation, 1919), 34, 38.

[46] C. C. Carstens, "How to Aid Deserted Wives," *Jewish Charity* 4 (February 1905): 1–40; Walter H. Liebman, "Some General Aspects of Family Desertion," *Journal of Social Hygiene* 1, no. 2 (April 1920); 205.

[47] Kate Holladay Claghorn, *The Immigrant's Day in Court* (New York: Arno Press and New York Times, 1969 [1923]), 82; Michael Francis Doyle, "Causes of Family Desertion," *City Club Bulletin (Philadelphia)* 6 (February 1913): 246.

[48] Charles Zusner, "Family Desertion (Report on a Study of 423 Cases)," *Annals of the American Academy* (1929): 101.

[49] Even when marriages are arranged, Jewish law requires that both parties to the marriage consent. Klein, *Religious Practice*, 390, 392. This does not preclude the possibility that some individuals may marry due to family pressure.

[50] riedman, "Husband Desertion," 3. However, Howe has argued that the transition to life in the USA was frequently easier for the wives, who adapted more quickly than their husbands. Howe, *World of Our Fathers*, 173.

[51] The protagonist Yekl in Abraham Cahan's novel *Yekl: A Tale of the New York Ghetto*, seeks recognition from his newly immigrated wife of his Americanization, represented by his "swell attire." Abraham Cahan, *Yekl: A Tale of the New York Ghetto* (New York: D. Appleton and Company, 1896), 33–34. For a more detailed discussion of Cahan's portrayal of Yekl's Americanization, see

> This was America, the Land of Freedom. No more oppressive religious obligations. No more chafing 'Don'ts.' Life must be lived to the fullest, material enjoyment must reign supreme ... The self-denials of the past were to be amply balanced by the self-indulgences of the present.[52]

Rabbino's perspective echoed that of M. Baranov who, decades earlier, had penned an article in *The Forward*, attributing men's desertions to their youth and their lack of spiritual roots:

> In Europe they were responsible for their lives; they lived within the framework of police regulations, religious ritual, teachings of relatives and neighbors. Every step was decided beforehand. Their road of life was narrow, but they could not get lost ... In America young Jews are hurled into a world of freedom—no fences, no police, no communal judgment. It's every man for himself. Nothing sacred; you can buy or sell everything for money. The aim of life is amusement; conscience and honor fall by the wayside ... Such a young man gets married. In three or four years he has several children, who are a nuisance. His wife grows sickly. His wages are too low to allow him any fun. He fights with his wife, who doesn't let him out of the house. There are gay young girls out there, and carefree bachelors. The anarchists preach free love; the freethinkers guarantee there is no God and no punishment in the afterlife. The young man thinks, "I'm a free person, who cares what they say," and one fine day he leaves home and forgets to come back. He becomes a missing husband.[53]

And religious obligations for Moische/Morris were many. It was never clear to me or any of Morris' other grandchildren whether he would have adhered to the many dictates of his Jewish faith without Gitla's/Gussie's needling and insistence. Gussie was relentless in her observance, praying numerous times each day. Like many immigrant Jewish women from Eastern Europe,[54] Gussie determinedly prepared for each Sabbath, cleaning their apartment to welcome the Sabbath bride, shopping on Jerome Avenue for the live fish that was to swim in their bathtub until it was time to meet its fate as gefilte fish, baking her special raisin-filled challah, lighting the Sabbath candles. Each Sabbath was bid farewell with the havdalah ceremony, complete with spices and the special intertwined candle.[55] The Sabbath traditions embued their home with a sense of warmth and comfort, but they also served as a reminder of one's religious obligations. Years later, Gussie's children Ruth and Robert and I and her other grandchildren would remember these rituals, but we

---

Cheryl Alexander Malcolm, "Othering the Other in Abraham Cahan's *Yekl: A Tale of the New York Ghetto*," *Amerikastudien/American Studies* 46, no. 2 (2001): 223–232.

[52] Bernhard Rabbino, *Back to the Home* (New York: Court Press, 1933), 29–30.

[53] M. Baranov, *Forward*, February 25, 1910. Quoted in Howe, *World of Our Fathers*, 180.

[54] Anne M. Polland, "The Sacredness of the Family": New York's Immigrant Jews and Their Religion, 1890–1930. PhD dissertation, Columbia University, 2004, 124–125.

[55] "Havdalah" signifies separation of the profane and the sacred. Just as the Sabbath is ushered in with the Kiddush ceremony, which begins with a blessing over wine, it is concluded with the Havdalah ceremony, which also begins with a blessing over wine. The Talmud teaches that each person receives a *neshamayetera,* an additional soul, on the Sabbath. The spices are said to soothe the soul of the person, who loses this additional soul at the Sabbath's conclusion. Klein, *Guide to Jewish Practice*, 73–74; "Havdalah: Transition, Separation, Memory." Accessed December 19, 2020. https://reformjudaism.org/havdalah-transition-separation-and-memory

grandchildren would have few memories of Morris wearing tefillin or a tallit or praying on a daily basis.[56]

Shortly before Gitla and Rivka arrived in New York City, Morris moved to the Mt. Eden area of the Bronx, variously considered to be a part of the South Bronx or the central Bronx,[57] to an apartment not far from Jerome Avenue, leaving Dora and their son Dale, born in 1926, in the Bensonhurst section of Brooklyn (see Figs. 2.7 and 2.8, depicting the various sections of the Bronx and Brooklyn, respectively). By 1930, Gitla had become Gussie, Rivka became Rose and then Ruth, and a second child was born to Morris and Gussie, Robert. We cannot know how each of them felt about the change in their names—whether this was a change to be embraced, a sign of their Americanization; or perhaps felt as a loss of some part of themselves, never to be retrieved; or merely a pragmatic modification that would allow them to more easily navigate their environment, without receiving stares from those who wondered at the foreignness of their names and their being.

Morris continued to work as a dressmaker as he had before Gussie's arrival. Although some of the building's other tenants had been born in New York, many were immigrants like themselves, having left their homelands in Germany, Poland, and Russia. The neighbors were employed in diverse occupations and industries: undertaking, millinery goods, interior decorating, life insurance, real estate, retail, and barber shops.[58]

It was likely a comfortable neighborhood for the family. In 1930, 57.1% of the South Bronx's population of 630,000 identified as Jews, with 62 different congregations, housed in a lesser number of synagogues.[59]

Bensonhurst, where Dora and Dale lived, sits in the southwestern part of Brooklyn, running from Stillwell Avenue and 86th Street at its southern tip, along Stillwell Avenue to Avenue P, east to McDonald Avenue, north to 60th Street,

---

[56] Tefillin are two small black boxes with attached straps that contain parchment scrolls bearing verses from the Torah. Jewish men are required to place one box at the center of their forehead and the second on their arm each weekday morning for prayer. There is a prescribed procedure, including prayers, for the donning and removal of the tefillin. The obligation to wear the tefillin derives from Deuteronomy 6:5–8. Klein, *Guide to Jewish Religious Practice*, 7–9; "Jewish Concepts: Tefillin," accessed December 18, 2020. https://www.jewishvirtuallibrary.org/tefillin. The tallit is a prayer shawl made from cotton, wool, or silk. Borderline fringes at its four corners, known as tzitzit, are knotted and twined. The tzitzit serve as a reminder of God's commandments and implore the wearer to obey these laws. The smaller version of the tallit, known as the tzitzit, is worn underneath or above one's clothing. The larger tallit, or tallit gadol, is worn on the outside of the clothing during prayer. Women are not required to wear a tallit or tzitzit and Orthodox Judaism forbids a woman to wear a tzitzit in public. Klein, *Guide to Jewish Religious Practice*, 3–6; "What Is A Tallit." *Jerusalem Post*, accessed December 18, 2020. https://wwww.jpost.com/special-content/what-is-a-tallit-631948

[57] Seymour J. Perlin, "Remembrance of Synagogues Past: The Lost Civilization of the Jewish South Bronx: Historical Survey," accessed December 13, 2020. http://www.bronxsynagogues.org/ic/bronxsyn/survey.html

[58] US Federal Census, 1930.

[59] Perlin, "Synagogues Past."

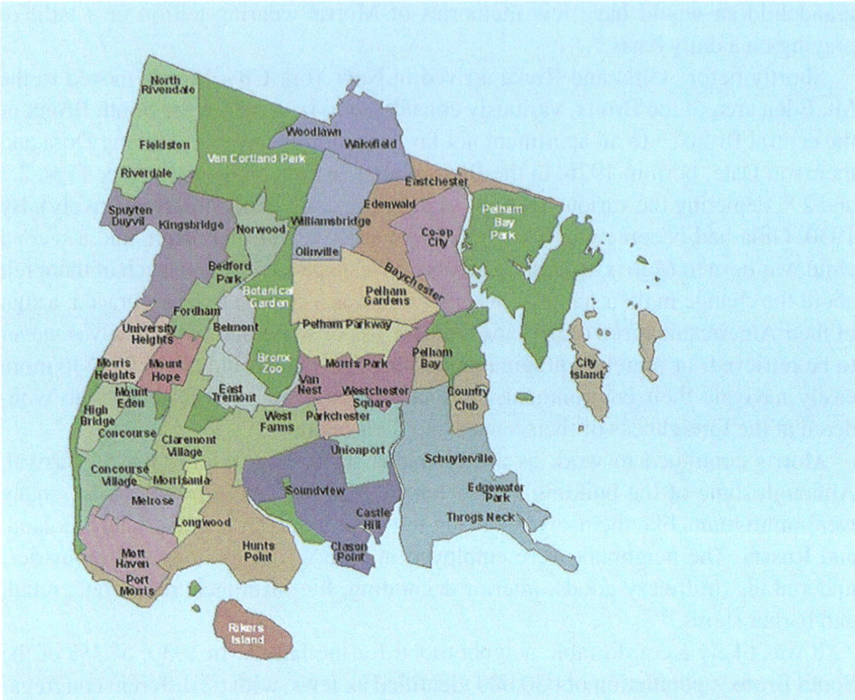

Fig. 2.7 Map entitled "New York City (NYC) Maps, Bronx with Community Districts" used with permission of the New York City Department of City Planning.

northwest to Fort Hamilton Parkway, southwest to Bay Ridge Avenue, southeast to 14th Avenue, south to 86th Street, and southeast back to Stillwell Avenue.[60]

The area had seen a steady increase in its population, particularly its Jewish and Italian population, since 1915, when the Fourth Avenue subway line, now the N and R trains, were brought into the area. By the time that Dora and Dale were living there in 1930, Bensonhurst had more than 150,000 residents, a Jewish Community House that contained a library, a swimming pool, bowling alleys, and classrooms,[61] and a number of synagogues.[62] Many of Bensonhurst's Jews, who had immigrated from Syria or Egypt, were Sephardic or Mizrahi, whereas the Jews from Poland and Russia, the birthplaces of Louis and Morris, were Ashkenazic.[63]

---

[60] Brooklyn Jewish Historical Initiative, "Bensonhurst," September 27, 2011, accessed December 3, 2020. https://brooklynjewish.org/neighborhoods/bensonhurst-brooklyn/

[61] "Jewish Centre Open Today," *New York Times*, March 4, 1928.

[62] Brooklyn Jewish Historical Initiative, "Bensonhurst"; "Bensonhurst's Past," accessed December 7, 2020. https://macaulay.cuny.edu/seminars/napoli08/articles/b/e/n/Bensonhurst%27s_Past_bc88.html

[63] Differences in Sephardic and Ashkenazic rituals largely emerged during the fifteenth and sixteenth centuries as a result of shifts in the Jewish population and the homogenizing effects of the

Fig. 2.8 Map entitled "New York City (NYC) Maps, Brooklyn with Community Districts" used with permission of the New York City Department of City Planning. All rights reserved

The Missing Safety Net: Poverty Unromanticized

It has been said that the story of Eastern European Jews is

> the greatest Horatio Alger story in American history. By the 1940s Jews in the USA (by now, of course, overwhelmingly Eastern European origins) had an occupation distribution which showed striking similarities to that of the high-status Protestants.[64]

---

printing press. The Yiddish language is common to Ashkenazic Jews and Ladino to Sephardic Jews, Adam Shear, "Jews and Judaism in Early Modern Europe," in *The Cambridge Guide to Jewish History, Religion, and Culture*, ed. Judith R. Baskin and Kenneth Seeskin (Cambridge: Cambridge University Press, 2010), 152–153.

[64] Milton Gordon, *Assimilation in American Life* (New York: Oxford University Press, 1964), 185. See Jesse E. Pope, *The Clothing Industry in New York* (Columbia, MO: University of Missouri, 1905), 47; Joel Seidman, *The Needle Trades* (New York: Farrar & Rinehart, 1942), 35–37; Jacob Riis, *How the Other Half Lives: Studies among the Tenements of New York* (New York: Charles Scribner's Sons, 1890), 88–89, 94. Abraham Lefcourt's life has been described as "an iconic Jewish American rags-to-riches story" that follows him from a Lower East Side tenement to success in the garment industry. Andrew S. Dolkart, "From the Rag Trade to Riches," in *Chosen*

Indeed, it has been argued that even during the Great Depression, 47% of New York's Jewish families were employed in white-collar occupations.[65]

But the Depression was not kind to Morris and Gussie and their children Ruth and Robert, and not to Dora, now essentially a single parent left to manage with her son Dale. Although it was not uncommon for Jewish women in Eastern Europe to work while their husbands spent their days studying Talmud and going to *shul* (synagogue),[66] this was not the life that Morris and Gussie lived. Morris would certainly never have permitted Gussie to work outside of the home; he would not even allow her to use the telephone once they had been able to acquire one. Like many men, Morris did not work full-time in 1939; his salary as a dress operator in a dress factory was similar to that of others in the clothing trades during the Depression era.[67] His work brought the family $1170, or $21,904 in today's dollars.[68] Ruth supplemented Morris' part-time income for the family by working as a floor girl in a dress factory, adding $832 to the family income that year, abandoning her dream of finishing college. Her employment was not unusual, even though she was still a teenager at the time. Teenage children of New York Jewish immigrant families often worked and their families frequently relied heavily on their wages.[69] It's hardly surprising that Morris did not have work each week of that year; the Depression had

Capital: The Jewish Encounter with American Capitalism, ed. Rebecca Kobrin (New Brunswick, NJ: Rutgers University Press, 2012), 86.

[65] Beth S. Wenger, *New York Jews and the Great Depression: Uncertain Promise* (Syracuse, NY: Syracuse University Press, 1999), 15.

[66] Nathan Hurvitz, "Courtship and Arranged Marriages among Eastern European Jews Prior to World War I as Depicted in a *Briefenshteller*," *Journal of Marriage and Family* 37, no. 2 (May 1975): 425. Zborowski explained a Briefenshteller as follows:

> And among the best sellers of all [books] is the *brifshteller*, the "letter-composer," with model letters to fit any occasion. Hundreds of women in hundreds of towns will write to their husbands identical letters, copied out of some *brifshteller*, each telling her husband about the special affection and esteem for him and about the health of their children. Those who are not literate enough to copy a letter from a book will employ the local letter writer to do it for them. When the answer comes, many a woman is unable to read it for herself, but must get some man to tell her what news it brings from her husband in America or her son in the service. She will beg him to go over it again and again until she knows every word. Then she will "read" it to her friends and neighbors, turning the page at the right place and point out, "See, where he says he misses his mammeh."

Zborowski, *Life with People*.

[67] In 1934, workers in what were considered Jewish industries, including the garment trade, were unable to find work on a regular basis and generally earned less than $1200 annually. Wenger, *Great Depression*.

[68] CPI Inflation Calculator.

[69] Polland, Sacredness of the Family, 119. A report of the U.S. Immigration Commission in 1910 indicated that, more than any other group, Russian Jews depended on their children's earnings. Russian Jewish children contributed an average of 30.7% to the household income, in comparison with children in other foreign-born families, whose wages accounted for 21.1% of the household income. US Congress, Senate, Reports of the Immigration Commission, vol. 1, "Immigrants in Cities," 61st Cong., 2d Sess. (Washington, D.C., 1911).

Fig. 2.9 Morris, Gussie, Ruth, and Robert, New York, circa 1937. (Source: Author)

brought massive layoffs to the garment industry[70] (see Fig. 2.9, for a photo of Morris, Gussie, Ruth, and Robert, during these difficult years. It is likely that Morris made the clothes seen in this photograph). Indeed, the manufacturing and industrial occupations employed slightly less than 30% of all workers in 1930, but by 1934, 53% of those who received unemployment relief had been employed in these sectors.[71] A 1937 report found that 12% of Jewish heads of households were unemployed and close to one-half of Jewish youths were as well.[72] Nevertheless, economists at the time considered unemployment to be not only a labor problem, but a moral problem as well.[73]

At the time, there was no federal provision for health care and workers were obliged to rely on their own funds and whatever funds might be available to them

[70] Joseph Brandes, "From Sweatshop to Stability: Jewish Labor Between Two World Wars," *YIVO Annual of Jewish Social Science* 16 (1976): 61–67.

[71] Gladys L. Palmer and Katherine D. Wood, *Urban Workers on Relief*(Washington, D.C.: Works Progress Administration, 1934), 6, 82.

[72] Nettie P. McGill, "Some Characteristics of Jewish Youth in New York City," *Jewish Social Service Quarterly* XIV (December 1937): 251–272.

[73] Michael B. Katz, *In the Shadow of the Poorhouse: A Social History of Welfare in America* (New York: Basic Books, 1996), 203.

from friends, family members, unions' reserve funds, and benevolent associations.[74] Even prior to the Depression, unions' reserve funds were overwhelmed by the number of unemployed individuals in the clothing, fur, and hat industries.[75] Funds became available through the states for both work relief and direct relief following the Congressional creation of the Federal Emergency Relief Administration in May 1933.[76] It was not until the passage of the Social Security Act in late 1935 that unemployment insurance became mandatory, and not until the 1940s and through collective bargaining that workers such as Morris were able to obtain health care benefits.[77]

Indeed, as an adult Ruth often recounted the difficulties that her family had faced during the Depression. Kosher meat was expensive,[78] and sometimes what was sold as kosher meat wasn't.[79] Like many US families who were compelled by necessity to change their eating habits,[80] Ruth's family shifted their diet to subsist on organ meats—liver, heart, and brain. The children's shoes were often too small, seeding the beginnings of the bunions that would cause Ruth's foot pain in later years. Perhaps it was these years that led Ruth in her adult years to maintain a closet full of shoes and freezers full of bread and meat for her family. As some scholars have noted, for many who lived during the Depression, "there will never be 'enough' security, money, or employment to make up for the devastating absence of these consumption essentials during their formative years."[81] And, perhaps Ruth's experience of poverty during this time may have contributed to what appeared to be untreated depression in her adulthood. Research findings now tell us of the association between experiences of childhood poverty and later adult depression.[82]

Ruth often spoke admiringly of her father, despite his refusal to allow her to attend college. Instead, she worked to help support the family, while her brother Robert (Bob) attended the university. She frequently voiced her resentment of

---

[74] See generally Kiran Klaus Patel, *The New Deal: A Global History* (Princeton, NJ: Princeton University Press, 2016).

[75] Daniel Nelson, *Unemployment Insurance: The American Experience 1915–1935* (Madison, WI: University of Wisconsin Press, 1969), 69, 79, 103.

[76] Katz, *Poorhouse*, 227.

[77] Katz, *Poorhouse*, 251.

[78] "Women's Strike Closes 5000 Meat Shops Here," *Daily News (New York)*, May 25, 1935; Guy Richards, "Meat Grading Seen as Strike Ends Here," *Daily News (New York)*, October 9, 1937.

[79] Gates Monroe, "2 Plead Guilty to Brooklyn Kosher Meat Fraud," *Daily News (New York)*, May 27, 1931; "Big Kosher Firm Guilty of Selling Non-Kosher Meat," *Daily News (New York)*, December 5, 1933.

[80] Janet Poppendeck, *Breadlines Knee-Deep in Wheat: Food Assistance in the Great Depression* (New Brunswick, NJ: Rutgers University Press, 1986); John A. Garraty, *The Great Depression* (New York: Harcourt Brace Jovanovich, 1986).

[81] Ronald Paul Hill, Elizabeth C. Hirshman, and John F. Bauman, "Consumer Survival During the Great Depression: Reports from the Field," *Journal of Macromarketing* (Spring 1997): 107.

[82] Lee et al., "Childhood Poverty"; Gary W. Evans, and Kimberly English, "The Environment of Poverty: Multiple Stressor Exposure, Psychophysiological Stress, and Socioemotional Adjustment," *Child Development* 73 (2002): 1238–1248.

Robert, who she felt, was insufficiently grateful for the sacrifice that she felt that she had made for him. Perhaps Morris' edict related not only to the financial situation, but also to Jewish tradition. According to Jewish law (*halakha*), women are exempt from the obligation to study the Hebrew Bible (*Torah*). However, as one scholar has noted, this became, in essence, a form of social exclusion.[83] Accordingly, it was the sons who brought honor to the family through their study and the sons who held privilege. Study at college was not study of the Torah, but sons had privilege, regardless.

Dora and Dale became almost untraceable through census records, moving from one boarding house to another, Dora cleaning offices at night to support her now three-year old son and relying on government assistance for rent and food.[84] Perhaps it was Flatbush's status as "Brooklyn's most affluent neighborhood in the Depression"[85] that provided Dora with these employment opportunities, but certainly nothing more than that. Dale thought that the people behind the large houses that he saw must be wealthy.[86] Although Morris had been making child support payments, he later fell behind with his payments and, in 1934 or 1935, found himself in court with Dora as a result. Some family members recall and relate this as the first meeting between Dale and his father, to be followed about a year later with a trip to the Bronx, where Dale met Gussie, Robert, and Ruth.[87] Others have framed the encounter somewhat differently, recalling instead that Dora and Dale appeared at Gussie's doorstep during the Depression, seeking support.

Either account comes to the listener, to the reader, long after the occurrence of the actual events, raising questions about each person's ability or willingness to remember events otherwise as they recall their past in what is now their present. Memory, after all, may be intentionally or unintentionally selective as individuals struggle toward a coherent understanding of themselves and their lives. There exists not only "the problematic of memory and faithfulness to the past; [but also] for forgiveness, guilt and reconciliation with the past [that is remembered]."[88]

> The person who seeks the connecting threads in the history of his life has already, from different points of view, created a coherence in that life which he is now putting into words. He has created it by experiencing values and realizing purposes in his life, making plans for it, seeing his past in terms of development and his future as the shaping of his life and of whatever he values most. He has, in his memory, singled out and accentuated the moments which he experienced as significant; others he has allowed to sink into forgetfulness. The

[83] Paula E. Hyman, *Gender and Assimilation in Modern Jewish History: The Roles and Representation of Women* (Seattle, WA: University of Washington Press, 1995), 54.

[84] Zand, *The Zands of Lowicz*, 19.

[85] Wenger, *Great Depression*, 89.

[86] Interview with Dale Zand, April 4, 2021.

[87] Interview with Dale Zand, April 4, 2021.

[88] Paul Ricoeur, *Memory, History, Forgetting*, trans. Kathleen Blamey and David Pellauer (Chicago, IL: University of Chicago Press, 2004), 412. One writer has suggested that "[o]ne is haunted by the past only when the past is remembered as radical difference." Michael P. Kramer, "Assimilation in the Promised Land: Mary Antin and the Jewish Origins of the American Self," *Prooftexts* 18, no. 2 (May 1998): 129.

> future has corrected his illusions about the significance of certain moments ... The units are formed by the conceptions of experience in which present and past events are held together by a common meaning. Among these experiences are those which have a special dignity both in themselves and for the course of his life; they have been preserved by memory and lifted out of the endless stream of what has happened and been forgotten.[89]

Nevertheless, Ruth and Robert never countered Dale's assertion that he rarely saw Morris, essentially growing up without a father, a circumstance that stands in sharp contrast with the myth of Jewish families, that '[h]uman relations are expected to endure."[90] In recounting the circumstances during this part of his life, Dale mused,

> I really was a problematic unwanted child. I was an outsider. I was causing a problem for Morris and his family, for Ruth and Bob and Morris. I was considered the great embarrassment and the great shame.[91]

Indeed, Dale's niece Peggy, my sister, born eight years after I, did not know of Dale's existence until the death of our father in 2006, when Dale appeared at his burial. Ruth had never mentioned her half-brother to her in the intervening 46 years between the time of Peggy's birth and the date of Henry's burial.[92] I have no memory of how I first learned about Dale but, given my relationship with Gussie, it was likely from Gussie herself.

References

1924 Immigration Act, Pub. L. 68–139, 43 Stat. 153 (1924).

Abrams, Nathan. "Digesting Woody: Food and Foodways in the Movies of Woody Allen," in *Woody on Rye: Jewishness in the Films and Plays of Woody Allen*, eds. Nathan Abrams, Vincent Brooks, and Marat Grinberg, 215–234. Waltham, MA: Brandeis University Press, 2014.

Act of Marriages A59/1918, City of Lowicz, Poland.

Akhtar, Salman. *Immigration and Identity: Turmoil, Treatment, and Transformation*. Northvale, NJ: Jason Aronson Inc., 1999.

Alroey, Gur. *Bread to Eat, Clothes to Wear: Letters from Jewish Migrants in the Early Twentieth Century*. Detroit, MI: Wayne State University Press, 2011.

Baranov, M. *Forward*, February 25, 1910.

Bier, Magdalena Ewa. *How to Become Jewish Americans?: The A Bintel Brief Advice Column in Abraham Cahan's Yiddish Forverts*. Frankfurt am Main, Germany: Peter Lang, 2015.

"Big Kosher Firm Guilty of Selling Non-Kosher Meat," *Daily News (New York)*, December 5, 1933.

Bitarin, Daniel and Boyarin, Jonathan. "Diaspora: Generation and Ground of Jewish Identity." *Critical Inquiry* 19 (1993): 693–725.

Brandes, Joseph. "From Sweatshop to Stability: Jewish Labor Between Two World Wars." *YIVO Annual of Jewish Social Science* 16 (1976): 61–67.

---

[89] H. P. Rickman, ed. *Meaning in History: W. Dilthey's Thoughts on History and Society* (London: George Allen & Unwin Ltd., 1961), 86–87.

[90] Zborowski, *Life with People*, 304.

[91] Interview with Dale Zand, April 4, 2021.

[92] Interview with Peggy Fogelman, March 31, 2024.

Brooklyn Jewish Historical Initiative, "Bensonhurst," September 27, 2011, accessed December 3, 2020. https://brooklynjewish.org/neighborhoods/bensonhurst-brooklyn/.
Brooklyn Jewish Historical Initiative, "Bensonhurst;" "Bensonhurst's Past," accessed December 3, 2020. https://macaulay.cuny.edu/seminars/napoli08/articles/b/e/n/Bensonhurst%27s_Past_bc88.html.
Burgess, Ann W. and Holstrom, Lynda L. "Adaptive Strategies in Recovery from Rape." *American Journal of Psychiatry* 136 (1979): 1278–1282.
Cahan, Abraham. *Yekl: A Tale of the New York Ghetto*. New York: D. Appleton and Company, 1896.
Carstens, C.C. "How to Aid Deserted Wives," *Jewish Charity* 4 (February 1905): 1–40.
Centers for Disease Control and Prevention, "Preventing Adverse Childhood Experiences," April 3, 2020, accessed December 14, 2020. https://www.cdc.gov/violenceprevention/aces/fastfact.html.
Claghorn, Kate Holladay. *The Immigrant's Day in Court*. New York: Arno Press and New York Times, 1969 [1923].
Colcord, Joanna C. *Broken Homes: A Study of Family Desertion and Its Social Treatment*. New York: Russell Sage Foundation, 1919.
Dolkart, Andrew S. "From the Rag Trade to Riches," in *Chosen Capital: The Jewish Encounter with American Capitalism*, ed. Rebecca Kobrin, 62–92. New Brunswick, NJ: Rutgers University Press, 2012.
Donim, Hayim Halevy. *To Be a Jew: A Guide to Jewish Observance in Contemporary Life*. New York: Basic Books, 1972.
Doyle, Michael Francis. "Causes of Family Desertion," *City Club Bulletin (Philadelphia)* 6, no. 11 (February 1913): 246–248.
Emergency Quota Act, Pub. L. 67–5, 42 Stat. 5 (1921).
Erikson, Erik. *Childhood and Society*. New York: W.W. Norton & Company, 1951.
Erikson, Erik. *Insight and Responsibility*. New York: W.W. Norton & Company, 1964.
Erikson, Erik. *The Life Cycle Completed (Extended Version)*.New York: W.W. Norton & Company, 1997.
Erikson, Erik. *Life History and the Historical Moment*. New York: W.W. Norton & Company, 1975.
Evans, Gary W., and English, Kimberly. "The Environment of Poverty: Multiple Stressor Exposure, Psychophysiological Stress, and Socioemotional Adjustment." *Child Development* 73 (2002): 1238–1248.
"Excerpt from Passenger Manifest, S.S. Berengia, Sailing from Cherbourg, France, arriving in New York March 29, 1929." Accessed https://heritage.statueofliberty.org
"Excerpt from Passenger Manifest, S.S. Ryndham, Sailing from Rotterdam, January 25, 1921." Accessed https://heritage.statueofliberty.org/
Finck, Liana. *A Bintel Brief: Love and Longing in Old New York*. New York: HarperCollins Books, 2014.
Fosha, Diana. "Dyadic Regulation and Experiential Work with Emotion and Relatedness in Trauma and Disordered Attachment," in *Healing Trauma: Attachment, Trauma, the Brain and the Mind*. Edited by Marion F. Solomon and Daniel J. Siegel, 221–281. New York: Norton, 203.
Friedman, Reena Sigman. 'Send Me My Husband Who Is in New York City': Husband Desertion in the American Jewish Immigrant Community 1900–1926." *Jewish Social Studies* 44, no. 1 (1982): 1–18.
Garaty, John A. *The Great Depression*. New York: Harcourt Brace Jovanovich, 1986.
Gordon, Milton. *Assimilation in American Life*. New York: Oxford University Press, 1964.
Grossberg, Lawrence. "Identity and Cultural Studies: Is That All There Is?" *Questions of Cultural Identity*. Edited by Stuart Hall and Paul du Gay, 87–107. London: Sage Publications, 1996.
"Havdalah: Transition, Separation, Memory." Accessed December 19, 2020. https://reformjudaism.org/havdalah-transition-separation-and-memory.
Hill, Ronald Paul, Hirshman, Elizabeth C., and Bauman, John F. "Consumer Survival During the Great Depression: Reports from the Field." *Journal of Macromarketing* (Spring 1997): 107–127.

Howe, Irving. *World of Our Fathers: The Journey of East European Jews to America and the Life They Found and Made*. New York: New York University Press, 2005.

Hurvitz, Nathan. "Courtship and Arranged Marriages among Eastern European Jews Prior to World War I as Depicted in a *Briefenshteller*." *Journal of Marriage and Family* 37, no. 2 (May 1975): 422–430.

Hyman, Paula E. *Gender and Assimilation in Modern Jewish History: The Roles and Representations of Women*. Seattle, WA: University of Washington Press, 1995.

Igra, Anna R. *Wives Without Husbands: Marriage, Desertion, & Welfare in New York, 1900–1935*. Chapel Hill, NC: University of North Carolina Press, 2007.

"Jewish Centre Open Today," *New York Times*, March 4, 1928.

"Jewish Concepts: Tefillin," accessed December 18, 2020. https://www.jewishvirtuallibrary.org/tefillin

Kalola, Bina. "Immigration Laws and the Immigrant Woman: 1885–1924." *Georgetown Immigration Law Journal* 11, no. 3 (Spring 1997): 553–600.

Katz, Michael B. *In the Shadow of the Poorhouse: A Social History of Welfare in America*. New York: Basic Books, 1996.

Klein, Isaac. *A Guide to Jewish Religious Practice*. New York: Jewish Theological Seminary of America, 1979.

Kramer, Michael P. "Assimilation in the Promised Land: Mary Antin and the Jewish Origins of the American Self." *Prooftexts* 18, no. 2 (May 1998): 121–148.

Lee, Hanna, Slack, Kristen S., Berger, Lawrence M., Mather, Rebecca S., and Murray, Rebecca K. "Childhood Poverty, Adverse Childhood Experiences, and Adult Health Outcomes." *Health & Social Work* 46, no. 3 (August 2021): 159–170.

Liebman, Walter H. "Some General Aspects of Family Desertion." *Journal of Social Hygiene* 1, no. 2 (April 1920): 197–212.

Malcolm, Cheryl Alexander. "Othering the Other in Abraham Cahan's *Yekl: A Tale of the New York Ghetto*." *Amerikastudien/American Studies* 46, no. 2 (2001): 223–232.

"Marketplace in Lowicz." By permission, from the Archives of The YIVO Institute for Jewish Research, New York.

McFarlane, Alexander C. and Yehuda, Rachel. "Resilience, Vulnerability, and the Course of Posttraumatic Reactions," in *Traumatic Stress: The Effects of Overwhelming Experience on Mind, Body, and Society*. Edited by Bessel A. van der Kolk, Alexander C. MacFarlane, and Lars Weisaeth, 155–181. New York: Guilford Press, 1996.

McGill, Nettie P. "Some Characteristics of Jewish Youth in New York City." *Jewish Social Service Quarterly* XIV (December 1937): 251–272.

Metzker, Isaac, ed. *A Bintel Brief, Vol. II: Letters to the Jewish Daily Forward, 1950–1980*, trans. Bella S. Metzker and Diana Shalet Levy. New York: Viking Press, 1981.

Metzker, Isaac, ed. *A Bintel Brief: Sixty Years of Letters from the Lower East Side to the Jewish Daily Forward*. New York: Doubleday & Company, Inc., 1971.

Monroe, Gates. "2 Plead Guilty to Brooklyn Kosher Meat Fraud," *Daily News (New York)*, May 27, 1931.

Nelson, Daniel. *Unemployment Insurance: The American Experience 1915–1935*.Madison, WI: University of Wisconsin Press, 1969.

New York City Department of City Planning "Map entitled 'New York City (NYC) Maps, Bronx with Community Districts'." Used with permission of the New York City Department of City Planning. All rights reserved.

New York City Department of City Planning "Map entitled 'New York City (NYC) Maps, Brooklyn with Community Districts'." Used with permission of the New York City Department of City Planning. All rights reserved.

Palmer, Gladys L. and Wood, Katherine D. *Urban Workers on Relief*. Washington, D.C.: Works Progress Administration, 1934.

Patel, Kiran Klaus. *The New Deal: A Global History*. Princeton, NJ: Princeton University Press, 2016.

Perlin, Seymour J. "Remembrance of Synagogues Past: The Lost Civilization of the Jewish South Bronx: Historical Survey." Accessed December 13, 2020. http://www.bronxsynagogies.org/'ic/bronxsyn/survey.html.
Polland, Anne M. "The Sacredness of the Family": New York's Immigrant Jews and Their Religion, 1890–1930. PhD dissertation, Columbia University, 2004.
Pope, Jesse E. *The Clothing Industry in New York*. Columbia, MO: University of Missouri, 1905.
Poppendeck, Janet. *Knee-Deep in Wheat: Food Assistance in the Great Depression*. New Brunswick, NJ: Rutgers University Press, 1986.
Prell, Riv-Ellen. *Fighting to Become Americans: Jews, Gender, and the Anxiety of Assimilation*. Boston, MA: Beacon Press, 1999.
Rabbino, Bernhard. *Back to the Home*. New York, Court Press, 1933.
Reiker, Patricia P. and Carmen, Elaine. "The Victim-to-Patient Process: The Disconfirmation and Transformation of Abuse." *American Journal of Orthopsychiatry* 56 (1986): 360–370.
Richards, Guy. "Meat Grading Seen as Strike Ends Here." *Daily News (New York)*, October 9, 1937.
Rickman, H.P., ed. *Meaning in History: W. Dilthey's Thoughts on History and Society*. London: George Allen & Unwin Ltd., 1961.
Ricoeur, Paul. *Memory, History, Forgetting*. Trans. Kathleen Blamey and David Pellauer. Chicago, IL: University of Chicago Press, 2004.
Riis, Jacob. *How the Other Half Lives: Studies among the Tenements of New York*. New York: Charles Scribner's Sons, 1890.
Salter, Anna C., *Transforming Trauma: A Guide to Understanding and Treating Adult Survivors of Child Sexual Abuse*. Thousand Oaks, CA: Sage, 1995.
Seidman, Joel. *The Needle Trades*. New York: Farrar & Rinehart, 1942.
Shaiak, G. "Lowicz: A Short History of the City and Its Jews," in *Lowicz: A Town in Mazovia Yizkor Book (Lowicz, Poland)*, ed. G. Shaiak. trans. from *Lowicz: ir be-Mazovia u-seviva, seferzikaron*. Tel Aviv, 1966. Accessed December 14, 2020. https://www.jewishgen.org/yizkor/lowicz/Lowicz.html#TOCE.
Shear, Adam. "Jews and Judaism in Early Modern Europe," in *The Cambridge Guide to Jewish History, Religion, and Culture*. Edited by Judith R. Baskin and Kenneth Seeskin, 140–168. Cambridge: Cambridge University Press, 2010.
Spicehandler, Miriam. "Market Day," in *Lowicz: A Town in Mazovia Yizkor Book (Lowicz, Poland)*. Edited by G. Shaiak. trans. from *Lowicz: ir be-Mazovia u-seviva,seferzikaron*. Tel Aviv, 1966. Accessed December 14, 2020. https://www.jewishgen.org/yizkor/lowicz/Lowicz.html#TOCE.
The Statue of Liberty—Ellis Island Liberty Foundation, Inc., https://heritage.statueofliberty.org/passenger-details/.
United States Congress, Senate, Reports of the Immigration Commission, vol. 1, "Immigrants in Cities," 61st Cong., 2d Sess. Washington, D.C., 1911.
United States Federal Census, 1930.
van der Hart, Onnon, Nijenhuis, Ellert R.S., and Steele, Kathy. *The Haunted Self: Structural Dissociation and the Treatment of Chronic Traumatization*. New York: W.W. Norton & Company, 2006.
van der Kolk, Bessel A. "The Complexity of Adaptation to Trauma: Self-Regulation, Stimulus Discrimination, and Characterological Development," in *Traumatic Stress: The Effects of Overwhelming Experience on Mind, Body, and Society*, eds. Bessel A. van der Kolk, Alexander C. MacFarlane, and Lars Weisaeth, 182–213. New York: Guilford Press, 1996.
van der Kolk, Bessel A. "The Compulsion to Repeat the Trauma: Re-enactment, Revictimization and Masochism." *Psychiatric Clinics of North America* 12, no. 2 (1989): 389–411.
van der Kolk, Bessel A. and McFarlane, Alexander C. "The Black Hole of Trauma," in *Traumatic Stress: The Effects of Overwhelming Experience on Mind, Body, and Society*, eds. Bessel A. van der Kolk, Alexander C. MacFarlane, and Lars Weisaeth, 3–23. New York: Guilford Press, 1996.
Wenger, Beth S. *New York Jews and the Great Depression: Uncertain Promise*. Syracuse, NY: Syracuse University Press, 1999.

"What Is A Tallit." *Jerusalem Post*, accessed December 18, 2020. https://jpost.com/special-content/what-is-a-talllit-631948.
Wheelis, Allen. *The Quest for Identity*. New York: W.W. Norton & Co., Inc., 1958.
"Women's Strike Closes 5,000 Meat Shops Here," *Daily News (New York),* May 25, 1935.
Zand, Martin. *The Zands of Lowicz: A Family History*. Rochester, NY: private printing, 2013.
Zborowski, Mark. *Life Is with People: The Jewish Little-Town of Eastern Europe*. New York: International University Press, 1952.

Lectures

Akhtar, Salman. "Shame." Presented at Cape Cod Institute, Eastham, MA, August 22, 2018.
Akhtar, Salman. Workshop "Fear, Greed, Shamelessness, Hatred, Guilt." Cape Cod Institute, Eastham, MA, August 20–24, 2018.

Interviews

Peggy Fogelman, March 30, 31, 2024
Dale Zand, April 4, 2021.

Chapter 3
The Present Past: Policy, Identity, and Progeny

Our memories are neither complete nor flawless. Rather, "[m]emory is partial, allusive, fragmentary, transient …."[1] Nevertheless, "[t]he past is never dead. It's not even past."[2] Our "communicative memory" embodies the knowledge of our past three generations, that we have been told directly by those who have had those experiences, with all of the narratives' flaws and imperfections.[3] When the memory across these generations encompasses trauma, the picture of that memory remains in the self and in the world becoming, in the words of Marianne Hirsch, rememory.[4] Those of us who have not directly experienced the trauma, but who carry its picture with us, become witnesses to the witness.[5] As such, we must strive to balance family narratives with external facts that have been documented and decipher our responsibility to our past in the present as we look to our future. As one cousin observed during my interview with them, "Stories are your truth until you know differently."[6]

My memories, understandings, and associations are assuredly not those of my parents or grandparents and are likely not precisely those of my siblings, although they may to varying degrees be congruent. These variations in memory are likely attributable to differences in age, sex, generation, birth order, sensitivities, and so forth. Additionally, because the events associated with some of my memories sometimes occurred prior to their births and their direct experience, my "communicative memories" are for them "institutionalized" or "cultural" memories; they must rely

[1] Kerwin Lee Klein, "On the Emergence of Memory in Historical Discourse," *Representations* Winter 2000, no. 69 (Winter 2000): 138.

[2] William Faulkner, *Requiem for a Nun*, reprint ed. (Knopf Doubleday Publishing Group, 2012).

[3] Marianne Hirsch, "The Generation of Postmemory: What, Who, When, Where, How?" Sponsored by the Museum of Jewish Heritage, New York, January 16, 2023 [virtual].

[4] Hirsch, "Postmemory," January 9, 2023.

[5] Hirsch, "Postmemory," January 23, 2023.

[6] Interview with Dror (pseudonym), United States, Spring 2024.

S. Loue, *From Public Policy to Family Dynamics*, SpringerBriefs in Social Policy, https://doi.org/10.1007/978-3-031-71818-2_3

on only family narrative for their understandings.[7] As with the previous chapters, what follows represents my effort to integrate and reconcile my memories, including sensory impressions, with documented events, to understand the impact of past policy on those who came before us and in our present.

Eugenic Policy, Unfulfilled Dreams, and Ambivalent Loss

Growing up it just seemed strange to me that Henry was so formal in his dress and his manner, always wearing slacks and a button-down shirt, most often a tie as well, even at home, never sporting a pair of jeans. Now as I look back I wonder if maybe he thought that this was what a doctor was supposed to look like, especially one that continued to make house calls to his patients long after other physicians had given up that practice, worried about being mugged for the drugs that they really didn't carry in their black bags or just too fatigued with the demands of their office practices and their long days to continue to see patients even after the dinner hour. Henry was always ready to drop everything at a moment's notice, his shoes on his feet even after he came home, even as he ate dinner, even as he sequestered in his study writing up his notes of the day, listening to one of his favorite symphonies—Mozart, Bach, Beethoven. This was my father's life, tending to the ills of his patients, so engrossed as he was in his passion for medicine and so firmly committed to the profession for which he had felt a calling since his earliest memories. There was little time left to attend to the demands of his five very different children, only to his wife who brought their every transgression to his attention for a pronouncement of punishment, however just or unjust it may have been. He had time and compassion for his children only when they were ill, transformed into patients. But for Henry, Michael's very existence appeared to be a transgression of incomprehensible enormity. He was not ill; he was abnormal, deficient.

A decade before Michael's birth in 1964, physicians most often recommended institutionalization for a retarded child.[8] Even at the time of his birth, children with Down syndrome were often still labeled "mongoloids," a term derived from the eugenic hierarchy, the attribution of the condition by John Langdon Down, whose name the condition bears, to the Mongolian "race,"[9] and the perpetuation and expansion of that notion by later scholars.[10] The children were then still assessed and

[7] Hirsch, "Postmemory," January 9, 16, 2023.

[8] Benjamin Spock, *Baby and Child Care* (New York: Basic Books, 1957); Abraham Levinson, *The Mentally Retarded Child* (New York: John Day, 1952).

[9] Daniel J. Kevles, "Mongolian 'Imbecility': Race and Its Rejections in the Understanding of Mental Disease," in *Mental Retardation in America: A Historical Reader*. Edited by Steven Noll and James W. Trent, Jr., 120–129 (New York: New York University Press, 2004).

[10] F. G. Crookshank, *The Mongol in Our Midst: A Study of Man and His Three Faces* (New York: E.P. Dutton, 1924). Crookshank maintained that "lower grade Mongolian imbeciles and idiots are undoubtedly orangoid [derived from orangutans] in their homologies," Crookshank, *Mongol*, 23.

categorized according to a classification system based on their putative intelligence level, one that had been developed in the early twentieth century: idiots, who would never attain a mental age of more than two years; imbeciles, whose intelligence was greater than that of an idiot, but who would never exceed the mental capabilities of a normal child of about seven years; and morons, with an intellectual development that exceeded that of an imbecile but was less than that of a normal 12-year-old child.[11]

Not surprisingly given this portrayal and the earlier association between mental retardation and eugenics, the very presence of the child in the home was perceived as a threat to the marriage and to the child's nondisabled siblings, who would experience a sense of family shame and be deprived of normal lives.[12] Physicians often assumed, as well, that both mothers and fathers would reject the child due to the child's intellectual deficiencies, "grotesque physical appearance," and the social stigma suffered by the family as a consequence of the child's existence.[13] Yet other professionals opined, "It becomes more difficult for the parent to give this affection since for most parents a retarded child represents both a lost hope for the future and a continuous and ever-repeated disappointment in the present."[14]

Henry had trained under a medical system, both in Switzerland and in the USA, that reflected these categorizations and labels, derived in large part from the eugenics movement.[15] The medical view of disability as a deficit inherent in the individual and the primary source of his or her identity, a model that reinforced the social separation and hierarchy of "normal" and "abnormal," had prevailed through the end of the 1920s, at which time the references to handicapped individuals replaced the language of defect.[16] The use of the term "handicapped" was sufficiently broad to encompass those deemed to be mentally retarded and rescue them from the label of "feeble-minded," with its associated legal, social, and medical consequences.

Fears of intellectually disabled persons' unhealthy impact on society through reproduction that had provided the justification for involuntary sterilizations began to give way by the 1940s to concerns for the healthy family members who would be

He further likened the hand lines of children with Down syndrome to those of the orangutan and characterized the brains of Down syndrome children as "simian." Crookshank, *Mongol*, 49, 73.

[11] Edmund Burke Huey, *Backward and Feeble-Minded Children* (Baltimore, MD: Warwick & York, 1912).

[12] C. Anderson Aldrich, "Preventive Medicine and Mongolism," *American Journal of Mental Deficiency* 52 (October 1947): 127–129.

[13] Harry Bakwin, "Informing the Parents of the Mentally Retarded Child," *Journal of Pediatrics* 49 (October 1956): 486–498.

[14] Arthur L. Rautman, "Society's First Responsibility to the Mentally Retarded," *American Journal of Mental Deficiency* 54 (1949–1950): 155–162.

[15] See Leila Zenderland, *Measuring Minds: Henry Herbert Goddard and the Origins of American Intelligence Testing* (New York: Cambridge University Press, 1998).

[16] Michael Grossberg, "From Feeble-Minded to Mentally Retarded: Child Protection and the Changing Place of Disabled Children in the Mid-Twentieth Century United States," *Paedagogica Historica* 47, no. 6 (2011): 729–747.

forced to bear the burdens of their mentally disabled family member.[17] Still, at least into the 1960s, both medical professionals and the general public referred to children with Down syndrome as Mongoloid idiots, imbeciles, or morons, depending upon their perceived intellectual capacity.[18]

Ruth and Henry could never have anticipated that they would have a fifth child, certainly not when Ruth was already beyond the age of 40.[19] Ruth confided years after Michael had been born that if abortion had been legal when she was pregnant with him, she would have likely opted for the procedure. Even so, antenatal screening for Down syndrome did not become possible until 1968, several years after Michael's birth.[20] Now, both amniocentesis and chorionic villus sampling are available to detect birth defects during the early stages of pregnancy, with the possibility of an abortion depending upon the test findings.[21]

Michael's appearance reflected the common features of individuals with Down syndrome—the shape of his eyes, the flat bridge of his nose, a tongue that was too large for his mouth, the positioning of the lines on his hands, and the various deformities of his toes—and this appearance seemed to irritate Henry to no end.[22] He

[17] Grossberg, "Feeble-Minded," 739.

[18] See, for example, Grossberg, "Feeble-Minded," 739; Alastair Beddie and Humphry Osmond, "Mothers, Mongols, and Mores," *Canadian Medical Association Journal* 73, no. 3 (August 1955): 167–170; L. Beckman, K.-H. Gustavson, and H.-O. Akesson, "Studies of Some Morphological Traits in Mental Defectives: Mongolism and Unspecified Mental Deficiency," *Hereditas* 48, no. 1–2 (1962): 105–122; Dennis Aldon Schmidt, "Review of Mongolism with Special Emphasis Upon Theories of Its Etiology (M.D thesis 2503," University of Nebraska Medical Center, 1960).

[19] It was noted during the early twentieth century that the birth of infants with Down syndrome seemed to have an association with the later age of the mother. Kevles, "Mongolian 'Imbecility'."

[20] N. J. Wald, A. Kennard, A. Hacksaw, and A. McGuire, "Antenatal Screening for Down Syndrome," *Journal of Medical Screening* 4, no. 4 (1997): 181–246.

[21] With chorionic villus sampling, a sample of tissue is taken from the umbilical cord. The advantage of this procedure over amniocentesis is that it can be done between 10 and 13 weeks of pregnancy, earlier than amniocentesis. However, the risk of miscarriage is slightly higher than with amniocentesis. American College of Obstetricians and Gynecologists, "Amniocentesis," April 2023. Accessed April 11, 2024, https://www.acog.org/womens-health/faqs/amniocentesis. The availability of post-screening abortion in Iceland has led to a significant decrease in the number of children with Down syndrome that are born each year. Julian Quinones and Arijeta Lajka, "What Kind of Society Do You Want to Live In? Inside the Country Where Down Syndrome Is Disappearing," CBS News, August 15, 2017. Accessed February 19, 2024, https://www.cbsnews.com/news/down-syndrome-iceland/. Following the 2022 U.S. Supreme Court's decision in *Dobbs v. Jackson Health Organization*, 597 U.S. _, women's ability to obtain an abortion of a child with Down syndrome following screening has been severely or completely restricted in many states. Amber Knight and Joshua Miller, "Afterword: Prenatal Genetic testing and Abortion After Dobbs," in *Prenatal Genetic Testing, Abortion, and Disability Justice*, 193–198. Eds. Amber Knight and Joshua Miller (New York: Oxford University Press, 2023).

[22] Deformities of the toes and feet are common among individuals with Down syndrome. Polly QX Lim, Nora Shields, Nikolaos Nikolopoulos, Joanna T. Barrett, Angela M. Evans, Nicholas F. Taylor, and Shannon E. Munteanu, "The Association of Foot Structure and Footwear Fit with Disability in Children and Adolescents with Down Syndrome," *Journal of Foot and Ankle Research* 8, no. 4 (2015): https://doi.org/10.1186/s13047-015-0062-0.

often scowled as he watched Michael attempting to feed himself, who often made quite the mess in the process as he struggled to coordinate the fork in his left hand. But it was Michael's mouth that invariably provoked Henry's anger, when Michael's tongue protruded from his not fully closed mouth, prompting Henry to swat him across the mouth as a mandate to withdraw his tongue and close his lips like a "normal" child. As Michael grew, Ruth said several times that she did "not want to look at his body." It was never clear whether she averted her gaze because of Michael's appearance or because of a sense of guilt or shame that she had produced him, or perhaps both.

Popular literature of the 1960s evidenced a slowly increasing number of testimonials from parents of retarded children that reassured parents that they could maintain their child at home, at least during their formative years and until their adolescence. A growing number of local associations of parents of retarded children were established at this time, as well as a national association.[23] Eunice Kennedy Shriver told the story of her sister, Rosemary, whose mild retardation following a severe bout of the flu later led to increasingly violent behavior that the family had hoped would subside following a lobotomy. Instead, the surgical procedure led to further mental disability, ultimately necessitating Rosemary's institutionalization. Shriver's *Saturday Evening Post* article assured the public that anyone could have a retarded child and that parents need not feel shame.[24] Henry and Ruth subscribed to the *Saturday Evening Post* for many years, but whether they saw this article, or whether they gave it any credence if they did read it, remains unknown.

How and why Ruth and Henry determined to bring Michael home has never been clear. There were loud arguments between the two and Ruth's pained sobs, particularly on and around the days when they were scheduled to drive Michael back to the institution from his brief two- or three-day visits with the family. Ruth commented years later that Michael seemed to be "getting worse" in the institution. Whatever animosities and jealousies existed between Ruth and her brother Robert (Bob), he signaled his respect for her for bringing Michael home. As he said to at least one of his children, it just wasn't done very often then.[25]

Policy and practice with respect to children with Down syndrome have undergone significant changes since the time of Michael's birth in 1964; the extent to which these changes ultimately influenced Ruth and Henry during the course of their lives with and decision-making regarding Michael remains unclear given the multiplicity of factors that they may have considered. The children are no longer referred to as mongoloids, Mongolian imbeciles, or retards, but as intellectually disabled individuals. The Arc, originally begun as the National Association for Retarded Children in 1953, advocated together with other organizations for the

[23] James W. Trent, *Inventing the Feeble Mind: A History of Intellectual Disability in the United States* (New York: Oxford University Press, 2017); Dale Evans Rogers, Angel Unaware (Fleming H. Revell Company, 1953); Pearl S. Buck, *The Child Who Never Grew* (New York: John Day, 1950).

[24] Eunice Kennedy Shriver, "Hope for Retarded Children," *The Saturday Evening Post*, September 22, 1962, 71–75.

[25] Interview with Dror (pseudonym), United States, Winter 2023.

establishment of the Medicaid and Medicare Programs, continuing with this advocacy effort into the 1970s, and was represented on the President's Panel for Mental Retardation.[26] A chapter of the organization once existed in Stamford, providing information for parents and a day program for young children. It was here that Michael first went to "school." Years later, Michael would have opportunities to participate in the Special Olympics, to attend a summer camp, and to acquire a greater range of daily living skills while attending a day program and a sheltered workshop.

At some point in Michael's twenties or thirties, Henry and Ruth placed Michael in the care of a local agency and had him moved to a somewhat independent living arrangement, in an apartment with a roommate and with support from staff. They seemed pleased and satisfied that Michael was now "living on his own." Their pleasure at his seeming accomplishment is not surprising in view of America's culture of independence and competition.[27] Although Michael had slowly acquired the skills that one would need for daily living—bathing, counting money, heating food—and despite the superficial appearance of normality, "living on his own" may well have been terrifying for him. During the short time when he was placed in his own apartment, albeit with some support, he picked at his arms and head until they bled, shaved off his eyebrows and sideburns, and scattered feces on the walls and floor of his bathroom. Neither the numerous medications that a psychiatrist had prescribed for his anxiety and depression nor Henry's frequent reprimands had much impact on the frequency of his self-mutilating behaviors.

Michael's behaviors should not have been surprising; professionals had noted even in the 1950s that mentally retarded children, as they were known then, who lacked security and acceptance in their own homes were more likely than "normal" children to develop behavioral difficulties.[28] Ruth and Henry seemed to be unaware of the living conditions within the apartment; they rarely, if ever, entered into his apartment, meeting him instead in the parking lot of the building when they wanted him to join them for an excursion. It was not until Michael began residing in a relatively well-staffed and supervised group home with the companionship of six or seven other intellectually and physically disabled adults that his self-inflicted holes on his arms and his head began to heal and his self-picking subside.

Although Ruth might have liked Michael to approximate the self-sufficiency of a "normal" child, she also seemed to be pleased that he continued to depend on her for affirmation and instrumental support. She proudly claimed that Michael was smart enough to not be taken advantage of and to have his own opinions, but she

---

[26] The Arc, "About Us, Our History." Accessed February 19, 2024, https://thearc.org/about-us/history/

[27] See Janice Brockley, "Rearing the Child Who Never Grew. Ideologies of Parenting and Intellectual Disability in the United States," in *Mental Retardation in America: A Historical Reader*. Edited by Steven Noll and James W. Trent, Jr., 130–164 (New York: New York University Press, 2004).

[28] Anne Marie Grebler, "Parental Attitudes Towards Mentally Retarded Children," *American Journal of Mental Deficiency* 55 (1951–1952): 476.

frequently rebuked him for not accomplishing a task quickly enough, such as cutting the steak on his plate, at which point she would take over and complete the task for him. Her caring, however ambivalent it may have been, served to maintain Michael's dependency on her and allowed her to retain her identity as a mother. Indeed, at the time of Michael's birth and beyond, societal expectation demanded that the mother of a disabled child assume primary responsibility for his or her training and development, while experts warned of the dangers to the family should the mother "become a martyr to a mentally retarded child."[29] Ruth may well have felt that she was in a societally constructed double bind—to provide sufficient care and training to her son to facilitate his maximally possible independence, but without devoting so much time to that function that she might neglect the other members of her family. This equivocal position may well have been exacerbated by feelings of guilt and sadness for having produced a "defective" child, for the impact of his birth on other family members, for Henry's unhappiness with Michael's birth and his hostility toward the child, sadness and guilt that were further compounded by Henry's belief that Ruth must have done something during her pregnancy to have brought about this *shanda* (שאַנדע; shame).

Eugenics, Antisemitism, and "Jewish Self-Hate"

I can close my eyes now and see Doris towering over us, more than six feet tall, always wearing a white uniform, white stockings, white shoes, as if she were really a nurse like she said, although we all knew that she wasn't. She also said she was like Mary Poppins, but we knew that that wasn't true either.[30]*Doris began staying with us each year from the time that I was about eight years old until I was about 15 or 16. She had been recommended to my parents by one of my mother's friends, and seemed to my parents to be a reliable caretaker, staying with us together with Gussie, while they went on their annual month-long visit to friends in Switzerland. Both Doris and my parents told us that Doris had come over to the United States from Germany at the close of the war and had been sponsored by a Jewish family somewhere in New York State. Morris had died a year or so before Doris began to come; my parents were concerned about leaving us with only Gussie and her ability to manage in an emergency, with her limited English, her even more limited telephone skills, and her total inability to drive a car if there were an emergency. It did not matter to Doris that she used the milekh (dairy) pots for meat, forcing Gussie to*

[29] Stella Stillson Slaughter, *The Mentally Retarded Child and His Parent* (New York: Harper & Brothers, 1960), 6–7. The feminist Betty Friedan observed that there were "mothers who devoted *too* much of their lives to their children, mothers who had to keep their children babies or they themselves would have no lives at all, mothers who never themselves reached or were encouraged to reach maturity …." Betty Friedan, *The Feminine Mystique* (New York: Laurel Books, 1984), 192.

[30] "Mary Poppins" was a musical fantasy comedy film produced in 1964 by Walt Disney. Julie Andrews played the role of Mary Poppins, the perfect nanny for two British children.

have to re-kasher the utensils and the pots to once again be able to use them.[31] *We told my parents when my sister was a year old, how Doris had tied her hands up when she could not control her scratching her chicken pox and later had left her naked in her high chair and tied her hands behind its back. We told my parents when Michael was perhaps all of three years old and now living at home, how Doris had dropped him into a bathtub of scalding hot water, hitting him on his back with her open palm in an effort to stop his screams, hitting him so hard that it left welts on his back. We told them how Doris tied our hands in back of our kitchen table chairs, daring us to eat our dinners without our hands like the "animals" that we were, how she threatened to punish us if we did not accompany her to Sunday mass at Sacred Heart, the Catholic church just down the street from where we lived. We told them how she bragged that her father had been a member of the Nazi Party, how she had believed in the tenets of the Nazi Youth Party.*[32] *They did not believe us, and Doris kept coming.*

Although Henry had grown up in a secular household, with apparently few, if any, ties to Jewish institutions or traditions, Ruth had been raised in an orthodox home that adhered to Jewish traditions and ritual observances, at least some of which she brought with her into her marriage with Henry. So it was a mystery to me, and a source of great anger, why my parents, who in those days kept kosher and attended synagogue, at least on the High Holy Days of Rosh Hashanah (the New Year) and Yom Kippur (the Day of Atonement), and sent us all to Hebrew school, had hired someone who seemed to so clearly hate Jews for being Jews.

As time moved on, and increasingly after the death of Morris, our house seemed to become less observant and, perhaps by some definitions, less Jewish and more assimilated. We moved from an ethnically diverse neighborhood with Jewish and Greek and Italian neighbors to a wealthier, predominantly Catholic area. As Christmas approached, we piled into my father's car at least once during that holiday season to see the Christmas lights that brightly adorned the homes and trees of nearby Christians. A Christmas tree stood in our family room each year, not a blue and white Chanukah bush as some might call it, but an actual Christmas tree. In accordance with orthodox Jewish teaching, I was not a bat mitzvah. After my brothers each had their bar mitzvah—the religious ceremony that recognizes a boy at the age of 13 as sufficiently mature to adhere to the religious precepts and participate in public worship—it seemed that our connections with the Hebrew school and synagogue dissolved, apart from my parents' payment of annual dues to their orthodox synagogue and my mother's attendance on Rosh Hashanah and Yom Kippur. The friendships that Henry and Ruth built and maintained for the greater part of their lives were with non-Jews. Once unthinkable, my father enjoyed ham sandwiches,

[31] The re-kashering of pots used for boiling can be done by boiling the pots for a prescribed amount of time. This readies the pots for cooking. The re-kashering of utensils can also be effectuated with boiling if the entire utensil is made of metal.

[32] Women were excluded from the Nazi Youth Movement because only men were believed to be fit to lead the Volk (Aryan people). Hyman, *Gender and Assimilation*, 141.

my mother ordered pork chops in restaurants, and we were all delighted when we could indulge in shrimp and lobster.

It has often been said that Jews face a double bind in that they wish to be accepted and assimilate into the larger culture, but continue to voluntarily segregate and maintain their cultural and religious distinctiveness, a conflict that may lead to a loss of self-esteem.[33] Perhaps this double bind explains Ruth's and Henry's gradual relinquishment of Jewish rituals and customs and their assumption of Christian-associated symbols. *But Doris?*

Some scholars have suggested that the era of the 1940s and 1950s was "an age of self-hatred" and that Jewish self-hatred was "'the neurosis' of the wartime generation."[34] It was during these years, in the late 1940s and early 1950s, that Henry attended medical school in Lausanne, Switzerland, and that Ruth worked for a large multinational corporation based in that country. Switzerland had a long history of both antisemitism and eugenics.[35] The voluntary segregation and unique traditions of Jews led European proponents of racial purity to accuse Jews of being too pure and therefore unable to assimilate; Jews were considered deviant because of their "race."[36] This deviancy or aberration was interpreted as a biological, and therefore hereditary, feature that was both irreversible and immutable.[37]

Switzerland was one of the last European nations to accord equal rights to Jews.[38] Prior to the First World War, Jews were blamed for the country's economic situation and, under the guise of protecting animals, a campaign was launched to prohibit Jews from engaging in the rituals governing the slaughter of animals, despite the intent of these rituals to protect the animals from fear and suffering.[39] The psychologist Carl Jung publicly declared that there existed a difference between Aryan and Jewish psychology.[40] The popular Swiss author and dramatist Friedrich Dürrenmatt, whose writings frequently addressed experiences of World War II and were widely read, asserted that an individual is not an antisemite even though he did not like

[33] Judith Weinstein Klein, "Ethnotherapy with Jews," *International Journal of Mental Health* 5, no. 2 (Summer 1976): 26, 28.

[34] Susan A. Glenn, "The Vogue of Jewish Self-Hatred in Post World War II America," *Jewish Social Studies* 12, no. 3 (Spring-Summer 2006): 98; Nathan A. Pelcovitz, "What About Jewish Anti-Semitism: A Prescription to Cure Self-Hatred," *Commentary* 3 (February 1947): 118.

[35] Caspar Battegay and Naomi Lubrich, Jewish Switzerland: 50 Objects Tell Their Stories/Jüdische Schweiz: 50 Onjekteerzählen Geschichte (Basel, Switzerland: Jüdisches Museum der Schweiz, Christoph Merian Verlag, 2018).

[36] David Mitchell and Sharon Snyder, "The Eugenic Atlantic: Race, Disability, and the Making of an International Eugenic Science, 1800–1945," *Disability & Society* 18, no. 7 (2003): 849–850.

[37] Mitchell and Snyder, "Eugenic Atlantic," 850; Zygmunt Bauman, "Modernity, Racism, Experimentation," in *Theories of Race and Racism: A Reader*, edited by Les Black and John Somos, 277–293 (New York: Routledge, 2000).

[38] Jacques Picard, *La Suisse et les juifs 1933–246; Anisemitisme Suisse, défense du judäisme, politique internationaleenvers les emigrants et les refugiés* (Lausanne, Switzerland: Editions d'en bas, 2000), 36.

[39] Picard, *La Suisse et les juifs*, 37–38.

[40] Picard, *La Suisse et les juifs*, 45.

Jews, and that the Jews themselves were responsible for the development of antisemitism.[41] During his residence in Switzerland, the Nobel Prize-winning novelist Thomas Mann was denied invitations to the bourgeois salons of Zurich because his wife, Katia Pringsheim, was born into a Jewish family,[42] In 1938, both the Swiss Department of Justice and the police justified their refusal to admit Jewish refugees beginning in 1933 by suggesting that the admission of Eastern European Jews into the country would foment antisemitism.[43] Indeed, the vast majority of people both inside and outside of Switzerland either never knew or have forgotten that the idea of marking Jews' passports with a "J" originated with the Swiss authorities.[44] In 1930, Jews accounted for 0.02 percent of all foreigners residing in Switzerland; by 1950, when Henry and Ruth were living in Lausanne, only 0.029% of foreigners were Jewish.[45] And, both during and after World War II, Switzerland served as safe repository for the Nazi wealth that they had plundered from their Jewish victims.[46]

And, despite Ruth's protestations that the Second World War and the Holocaust had "nothing to do with [them]," Henry and Ruth experienced Swiss antisemitism directly during there lives there. On one occasion, not long after settling in Lausanne, their tranquil restaurant dinner was abruptly interrupted by the intentionally loud comment of restaurant clients seated at the next table, who opined while gazing directly at Henry and Ruth, "Hitler should have finished the Jews off."

In the context of German Jewry, centuries of antisemitism, the idealization of "Germanness," and Jewish efforts to assimilate were thought to have produced the psychological syndrome of Jewish self-hatred.[47] The social psychologist Kurt Lewin posited that the antisemitism experienced by German Jews had led to a sense of shame among entire generations of Jews.[48] One might wonder whether the

---

[41] Picard, *La Suisse et les juifs*, 43.

[42] Picard, *La Suisse et les juifs*, 43.

[43] Picard, *La Suisse et les juifs*, 39.

[44] Battegay and Lubrich, *Jewish Switzerland*, 160.

[45] Picard, *La Suisse et les juifs*, 61 Table 1.

[46] Peter Capella, "Swiss Found Culpable of 'Helping Nazis'," *The Guardian*, December 10, 1999. Accessed March 29, 2024, https://www.theguardian.com/world/1999/dec/11/1. A report of the independent Bergier Commission concluded that "Swiss actions were influenced by latent, though nonviolent, anti-semitism that was 'deeply ingrained' in local society at the time" and that the Swiss had morally capitulated to the Nazis.

[47] See Todd M. Endelman, "Jewish Self-Hatred in Britain and Germany," in Two Nations: British and German Jews in Comparative Perspective, ed. Michael Brenner, Rainer Liedtke, and David Rechter (Tubingen, Germany: Mohr Siebeck, 1999), 333, 345; Sander L. Gilman, Jewish Self-Hatred: Anti-Semitism and the Hidden Language of the Jews (Baltimore, MD: Johns Hopkins University Press, 1986), 248, 293–295, 301–304. Glen has attributed the development of this concept to three primary influences: the increasing influence of psychological experts on American thinking during the post-World War II period; the transfer of European perspectives relation to Jewish self-hatred, Jewish self-consciousness, and antisemitism to the U.S. by Jewish intellectual immigrants; and what she has termed the "Jewish Cold War," which involved significant debate within the U.S. Jewish community with respect to Jewish identity. Glenn, "Vogue," 99–100.

[48] Kurt Lewin, "Everything Within Me Rebels: A Letter from Kurt Lewin to Wolfgang Kohler, 1933" trans. Miriam Lewin and Gabriel Wickert, *Journal of Social Issues* 42 (1986): 39–47.

longstanding antisemitism in Switzerland instilled this same sense among Swiss Jews, or whether the close proximity of Switzerland's German-speaking cantons to Germany may have facilitated their development of similar sentiments. Lewin asserted further that such self-hatred was often subtle and its existence and manifestations intertwined with other motivations, rendering it difficult to determine whether self-hatred was present.[49] Jewish self-hatred could be directed against oneself, against all Jews as a group, or against any aspect of Jewish life. Other mental health professionals suggested on the basis of their examination of psychiatric experts that Jewish self-hatred was a defense mechanism that developed from children's feelings of rejection, inferiority, and weakness stemming from their interactions with their overbearing parents.[50]

In the USA, the label of "a self-hating Jew" was later directed at Jews who questioned the need for a Jewish state, Jews who did not maintain ties to Jewish institutions or prefer to associate with Jews, and Jews who opposed the integration of African Americans into Jewish schools and residential areas because it would threaten the voluntary Jewish self-segregation that had allowed Jewish values and customs to survive.[51] Others contended that self-hate is the result not of being a Jew but of the discrimination, threats, and fear of violence which are present in the social situation in which the Jew finds himself….Neurotic Jews are such because of anxieties, stemming in part outside their group. To become a member of a Jewish communal organization does not remove the source of anxiety. At best such a procedure is a partial palliative which attempts to meet an evil condition with an inflation of the ego.[52]

A recent study appears to confirm this supposition. Researchers found that of the 242 Jewish-identified individuals participating in the study, those who had felt endangered because of their Jewish identity had higher rates of depression and self-hate, including internalized antisemitism.[53] However, whether these current-day

---

[49] Kurt Lewin, "Self-Hatred Among Jews," *Contemporary Jewish Record* 4, no. 3 (June 1941): 219–232.

[50] Nathan Ackerman and Marie Jahoda, *Anti-Semitism and Emotional Disorder: A Psychoanalytic Interpretation* (New York, Harpers, 1950). See also Nathan Ackerman, and Marie Jahoda, "The Dynamic Basis of Anti-Semitic Attitudes," *Psychoanalytic Quarterly* 17 (1948): 248, 258–260.

[51] Mordecai M. Kaplan, The Future of the American Jew (New York: Macmillan Company, 1948); Nathan Glazer, "Negroes and the Jews: The New Challenge to Pluralism," *Commentary* 38 (December 1964): 29–34. See also Seth Forman, *Blacks in the Jewish Mind: A Crisis of Liberalism* (New York: New York University Press, 2000), 59–64. Jewish ethnotherapy was developed in the 1970s as a technique to help Jews confront their internalized antisemitism, the existence of which may have manifested as opposition to the Israeli state, noninvolvement with Jewish institutions, and dissatisfaction with one's body image. Klein, "Ethnotherapy with Jews," 26–38. See also Nicholas Behrman, "The Therapy Framework of Jewish Ethnotherapy." Accessed January 4, 2024, https://kinkedin.com/pulse/therapy-framework-jewish-ethnotherapy-nicolas-behrmann-rabbi-pmp/

[52] L. O. Kattsoff, "Editorial: Definition of a Situation: The Jew in America," *Social Forces* 27, no. 4 (May 1949): 463.

[53] Sari M. Kosdon, Alex Keller, Jack W. Berry, David J. Stiver, Kelly Whiteford, and Lynn E. O'Connor, "The Experience of Being Jewish: Living with Antisemitism," *PSOCIAL* 7, NO. 1 (January–June 2021): ISSN:2422-619X.

research findings can be applied to those living during an earlier historical period is questionable.

Whether Henry or Ruth, my parents, experienced self-hatred is unknowable now. It seemed that they were, however, consumed with shame. Feelings of shame may arise in relation to elements of a specific circumstance, one's behavior or oneself that one considers to have fallen short of moral, aesthetic, or performance standards or that are viewed negatively.[54] Unlike guilt, which is oriented toward others, shame is thought to be oriented to the self; it is a violation of one's own ideals, rather than a violation of an externally imposed prohibition or standard.[55]

Whether the shame that Ruth experienced derived from instances of abuse or poverty or years' abandonment by her father or antisemitism or a lack of formal education or any combination of these and perhaps more remains unknown. Henry, too, seemed to embody shame, despite his professional accomplishments. Perhaps it stemmed from his experiences of antisemitism or the birth of his "defective" son Michael or his father's coarse and abrasive behavior towards the world or his father's verbal abuse of his mother, or his father's pronouncements that he was "worthless," or his mother's quiet victimization. Whether Henry's decision to study medicine in Switzerland was a result of the barriers created by medical school admission quotas for Jews that existed at the time of his application for admission or subpar undergraduate grades or both remains unknown today,[56] but either scenario may have inculcated a sense of shame associated with who he was or what he had done.

If either Ruth or Henry were alive today and were asked about the source of their shame, they would likely deny ever feeling that emotion. Research tells us now that repeated childhood experiences of shaming in connection with feelings, competencies, and needs may led to an internalization of shame, such that the child develops a sense of unworthiness and inferiority that continues into adulthood and becomes part of their identity, although it may not be recognized as such.[57] Too, acute, painful experiences may cause a sense of paralysis and, when repeated over time, may lead to a pervasive belief in one's inadequacy across various domains of functioning.[58]

[54] Dawn Leeming and Mary Boyle, "Shame as a Social Phenomenon: A Critical Analysis of the Concept of Dispositional Shame," *Psychology and Psychotherapy: Theory Research, and Practice* 17 (2004): 375.

[55] Fabrice Teroni, and Julien A. Deona, "Differentiating Shame from Guilt," *Consciousness and Cognition* 17, no. 3 (September 2008): 725–740.

[56] Many medical schools established quotas limiting the number of Jews who could be admitted. The practice appears to have begun in the 1920s and was firmly established in many schools by 1945. See Edward C. Halperin, "Why Did the United States Medical School Admissions Quota for Jews End?" *American Journal of the Medical Sciences* 358, no. 5 (2019): 317–325; Leon Sokoloff, "The Rise and Decline of the Jewish Quota in Medical School Admissions," *Bulletin of the New YorkAcademy of Medicine* 68, no. 4 (November 1992): 497–518.

[57] Gershen Kaufman, *The Psychology of Shame: Theory and Treatment of Shame-Based Syndromes* (New York: Springer, 1989).

[58] Janice Lindsay-Hartz, "Contrasting Experiences of Shame and Guilt," *American Behavioural Scientist* 27 (1984): 689–704.

These painful experiences may have been their own, or they may have been memory as well, as they carried the picture of their parents' and grandparents' experiences in themselves, to be compounded with their own.

Ruth's and Henry's self-denial of shame and experiences that likely constitute the source of that feeling may well have led to their later forgetting of these experiences, despite the ongoing sense of shame. Current understandings of denial indicate that a denial is a coping strategy that is often used to deal with experiences of trauma and associated feelings of shame and/or external pressures,[59] and self-generated denials of an event's details leads to adverse effects on memory, a phenomenon that scholars have termed denial-induced forgetting.[60] Too, their continuing denials of the truth of or accounts of Doris' behavior allowed them to suppress unwanted, painful memories and emotions that would have entered into their awareness.[61] Shame is often a reflection of one's perceived power or status relative to those around them.[62] Ruth's and Henry's feelings of shame may have been particularly painful because the underlying experiences that gave rise to them were beyond their control: their own and their parents' experiences of and efforts to survive legally condoned antisemitism in their countries of origin and the USA and social ostracism and disdain because of their Jewishness.

Antisemitism, too, created a distance from the larger surrounding society, a feeling of insecurity and, ultimately, a sense of isolation and loneliness for Louis and Annie, Morris and Gussie, Henry and Ruth, Dale and perhaps Dora as well. As yet another example, believing that his birth name of Isadore Ezra too easily signaled his Jewishness to others and would foreclose employment, economic, and social opportunities, Dale in 1948 legally changed his name to Dale Ezra.[63]

The impact of these multi-generational experiences has not spared my generation. One sibling, who shall remain unnamed, struggled for years to locate his

---

[59] Azadé Azad and Lina Leander, "Children's Reporting about Sexual Versus Physical Abuse: Patterns of Reporting, Avoidance and Denial," Psychiatry, Psychology, and Law 22 (2015): 890–902; Roy F. Baumeister, Karen Dale, and Kristin L. Sommer, "Freudian Defense Mechanisms and Empirical Findings in Modern Social Psychology: Reaction Formation, Projection, Displacement, Undoing, Isolation, Sublimation, ad Denial," *Journal of Personality* 66 (1998): 1081–1124.

[60] Fabina Battista, Antonietta Curci, Ivan Mangiulli, and Henry Otgaar, "What Can We Remember After Complex denials? The Impact of False Denial on Memory," *Psychology, Crime and Law* 2021: https://doi.org/10.1080/1068316X.2020.1865956; Henry Otgaar, Mark L. Howe, Tom Smeets, and Jianqing Wang, "Denial-Induced Forgetting: False Denials Undermine Memory, But External Denials Undermine Belief," *Journal of Applied Research in Memory and Cognition* 5, no. 2 (2016): 168–175; Henry Otgaar, Mark L. Howe, Amina Memon, and Jianqing Wang, "The Development of Differential Mnemonic Effects of False Denials and Forced Confabulations," *Behavioural Sciences and the Law* 32, no. 6 (2014): 718–731; Tameka Romeo, Henry Otgaar, Tom Smeets, Sara Lanström, and Didi Boerboom, "The Impact of Lying About a Traumatic Virtual Reality Experience in Memory," *Memory & Cognition* 47, no. 3 (2019): 485–495.

[61] Henry Otgaar, Tameka Romeo, Nikki Ramakers, and Mark L. Howe, "Forgetting Having Denied: The 'Amnesic' Consequences of Denial," *Memory & Cognition* 46 (2018): 520–529.

[62] Leeming and Boyle, "Shame Analysis," 383.

[63] Interview with Dale Zand, October 7, 2023.

identity, first proudly displaying Nazi paraphernalia on his bedroom walls while in high school, later becoming a believer in the rhetoric of Meier Kahane, the American-born Israeli extremist and cofounder of the Jewish Defense League and, later still, converting to Roman Catholicism.[64] Another sibling has experienced prolonged periods of severe mental distress, rooted in part in painful feelings of shame. I changed my name legally in my early twenties, feeling that I had been wearing my birth name as if it were a heavy hand-me-down coat, too big at the shoulders and swirling below my knees, a coat so large that it would never fit and that was at odds with my identity. Like Morris' efforts to wage battle against the injustices then-prevalent within the garment trades, I too have rebelled against injustices—those meted down by my parents, by Doris and, to the extent that I have been able to do so, by the larger society.

Poverty, Shame, Exclusion, and Isolation

My mother Ruth used to delight in telling me the story of her brother and the eggs. They were walking home from school and, along the way, cartons of eggs had fallen from a truck. Her brother Robert (Bob) stopped to collect some of the unbroken eggs in his coat pockets, knowing that they would be of help to their mother with all of her baking. Some other boys saw him put the eggs in his pockets and quickly rushed up, smashing them. He arrived at home with egg dripping from his pockets. My cousin tells me now that he was like that throughout his life, always bringing home things that were free, thinking that they might one day be needed. My cousin says that he always said that he never felt poor even though they were poor, that he saw the glass half-full, that here was something that perhaps could be useful. I wonder, though, why someone would acquire such varied objects with no immediate need or purpose for them, except to perhaps fill a hole of which one might not even be aware. I remember my mother, always keeping a freezer full of bread, her closet full with the many boxes of well-fitting dresses and designer shoes, all of which she lacked growing up.

[64] The Jewish Defense League (JDL) was founded in New York City in 1968, with its proclaimed purpose to protect Jews from local antisemitism. The JDL rested on five principles: pride and dignity, a love of Jewry, discipline and unity, and "iron," that is, the need to support Jews everywhere and to change the image of Jews by whatever means necessary, and a belief in the indestructibility of the Jewish people, Kahane encouraged Jews to take up weapons against those who oppressed them. The FBL classified the JDL as a right-wing terrorist organization in 2001, and it has been designated as a hate group by the Southern Poverty Law Center. Anti-Defamation League, "Backgrounder: The Jewish Defense League," April 14, 2010. Accessed March 29, 2024, http://www.adl.org/extremism/jdl_chron.asp; United States Department of Justice, Federal Bureau of Investigation, "Terrorism 2000/2001," October 7, 2012. Accessed March 29, 2024, https://web.archive.org/web/20121007083445/ https://www.fbi.gov/stats-services/publications/terror; Libby Kahane, *Rabbi Meir Kahane: His Life and Thought, Volume 1, 1932–1975* (CreateSpace Independent Publishing Platform, 2013).

From the age of one year until he was 16–1/2, Dale and his mother resided in boarding houses located in various sections of Brooklyn. Between the ages of one and seven, he and his mother boarded with a family that had three daughters and three sons. The sons lived on the third floor, two daughters lived on the second, and one daughter, who was married, lived elsewhere. Dale's mother was taken to the hospital. He recounts that at the age of three, "I was left alone. I cried incessantly for two nights. The older daughter yelled from the front room, 'Will someone shut that brat up!' The younger daughter ... she took me into her bed and held me until I stopped crying. And that was the beginning of feeling grateful for support." Dale remembered a time, as well, when he was eight years old and living at a different boarding house. He muses, "There was a big porch around the house and guys were playing baseball next door. I was playing by myself, making believe I was in a rocket, on another planet, like kids do. Then I stopped. It was as if I was outside of myself looking down. And I realized, I am all alone in the world. The only one who is going to take care of me is me. And the most important thing is to take care of myself." When asked about the effect that the relationship between his father Morris and his mother Dora may have had on him, Dale mused, "Perhaps subconsciously I felt an obligation to take care of the family, to care for the other person, a commitment not to be breached. I am not sure if the children realized it, but I tried my best to be tolerant and to give them support."

Andrew R. Heinze has suggested that European Jews who immigrated to the United States between 1881 and 1920 "moved from a world of scarcity to a world of abundance" and adapted to consumer attitudes in the USA.[65] Perhaps, however, this "abundance" represented more than an adaptation to consumer attitudes. In his adult life, Ruth's brother Bob was known to shop for sale meat at Wholesome Foods once each month, storing it in his family's basement freezer that he had brought home from his lab when it was no longer used there. He would bring home on a regular basis any leftover cookies and fruit cakes that were served at functions he attended, certain that someone at home would want to eat them.[66] For Ruth, the many boxes of shoes in her closets and the seemingly innumerable loaves of bread and chicken parts in the freezer seemed to signify some kind of insurance and assurance that she would not again face the poverty that she had experienced as a small child in Poland or during the US Depression. Perhaps, more importantly for Ruth, these goods provided her with the outward signs of being American and a member of the well-to-do middle class—she had, after all, denied for decades that she had been born in Poland.

Perhaps their move to the suburbs, to "north of the Parkway," provided both Ruth and Henry with additional assurance that they were no longer poor, that they would not be poor, and, for Ruth, that she had truly become American. Suburbanization has often served as a mechanism for Jews' incorporation into "America's

---

[65] Andrew R. Heinze, *Adapting to Abundance: Jewish Immigrants, Mass Consumption, and the Search for American Identity* (New York: Columbia University Press, 1990), 219.

[66] Interview with Dror (pseudonym), United States, Spring 2024.

socioeconomic mainstream."[67] Henry and Ruth and their then four children moved to the suburbs in 1964, leaving their small house and their Greek and Italian and Jewish neighbors. In some sense, Ruth's and Henry's ability to build this home in a majority Catholic suburb in the 1960s represented a significant shift from the days when a mortgage loan to Jews was considered especially risky and when restrictive covenants barred homeowners from selling to Jews or those who were not Caucasian, a racial classification from which Jews were excluded.[68] The new split-level that they had so carefully designed and watched as it came into being sat on an acre or so of land. The landscapers kept the lawn and gardens immaculately groomed and the housekeeper and periodic au pairs did the same with the interior. The walls of the hallways, covered with a variety of rich fabrics, ranging from silk to bamboo, were adorned with paintings and lithographs from Henry's art collection, which held pieces that were later borrowed by museums for their exhibits. Despite their middle-class status, despite their travels to Europe, despite their acquisition of luxuries that may have once seemed unattainable, Ruth in particular never displayed the level of comfort or assurance speaking with physicians, lawyers, teachers, or other professionals as she did with those who more closely reflected workers in the world from which she had come and had been raised—butchers, shopkeepers, and tailors.

Even in the USA, in this "world of abundance,"[69] Ruth and Robert and Dale, like their parents before them, had felt and suffered the effects of material scarcity—a lack of adequate clothing, the emptiness associated with insufficient, nutritious food, a situation now referred to as "food insecurity."[70] The wages that Morris had earned in the garment industry were inadequate to support two families, even if he had wanted to do so, and the lack of a government-established safety net in times of underemployment or unemployment rendered it impossible. Their poverty resulted

[67] William E. Nelson and Norman R. Williams, "Suburbanization and Market Failure: An Analysis of Government Policies Promoting Suburban Growth and Ethnic Assimilation," *Fordham Law Journal* 27, no. 1 (October 1999): 199.

[68] For a discussion of the practice of redlining that excluded Jews from better neighborhoods, see Beryl Satter, *Family Properties: Race, Real Estate, and the Exploitation of Black Urban America* (New York: Macmillan, 2009), 42; Ryan Reft, "Segregation in the City of Angels: A 1939 Map of Housing Inequality in L.A.," KCET website, January 19, 2017. Accessed January 15, 2024, https://kcet.org/shows/lost-la/segregation-in-the-city-ofangels-a-1939-map-of-housing-inequality-in-la. For a discussion of Jews' inability to purchase property due to restrictive covenants, see Nelson and Williams, "Suburbanization," 214, 222; John P. Dean, "Only Caucasian: A Study of Race Covenants," *Journal of Land & Public Utility Economics* 23, no. 4 (November 1947): 428–432; Michael Selzer, ed. "Kike!": A Documentary History of Anti-Semitism in America (New York: World Publishing, 1972), 205; Mary Ellen Stratthaus, "Flaw in the Jewel: History of Discrimination Against Jews in la Jolla, California," *American Jewish History* 84, no. 3 (September 1996): 189–219.

[69] Heinze, *Adapting to Abundance*, 219.

[70] Food insecurity is currently understood to be "a household-level economic and social condition of limited or uncertain access to adequate food." United States Department of Agriculture, Economic Research Service, "Definitions of Food Security," October 25, 2023. Accessed March 29, 2024, https://www.ers.usda.gov/topics/food-nutrition-assistance/food-security-in-the-u-s/definitions-of-food-security/

not only in economic exclusion, but social exclusion as well, due to their inability to afford the garments, transportation costs, and entrance fees that would have allowed them to participate in outings and events with their peers.

Members of my generation carry these multi-generational experiences of poverty with them. My sister Peggy has confided that, to this day, she has a deep fear of dying alone and penniless, perhaps stemming from the poverty that Ruth and Gussie experienced over their lifetimes.[71] My cousin Dror (pseudonym) tells me that he/she hears the voice of their parent in their head, retelling the story of their poverty, instilling in my cousin a sense of guilt whenever their family spends money on a nanny or household help.[72] Even if my cousin had known nothing of the parent's earlier poverty story, my cousin could see the effects of the parent's dire childhood poverty, the parent's constant efforts to defend against imagined slights, their seemingly incessant competition to better those who seemed to threaten their status, all were sufficient to let Dror know that the parent's childhood had been much less than ideal.

Welfare policy and immigration policy, now as then, are inextricably intertwined. At the time that Morris immigrated to the USA, immigration policy and law provided that individuals who would likely be unable to support themselves should be denied entry into the country.[73] Now, immigration law provides for the inadmissibility of persons who are "likely to become a public charge," affecting not only individuals seeking admission into the United States as tourists, or students, or for permanent residence, but affecting even individuals who are already permanent residents ("green card" holders) and who have left and are returning to the U.S. under specified circumstances.[74] Additionally, most immigrants, including those who are lawfully in the USA, are considered to be ineligible for most forms of publicly funded programs, including social security and Medicaid, unless they have been permanent residents for a minimum period of five years or meet other requirements.[75] Just as there was no safety net for Morris and Gussie and Ruth and Robert, Dora, and Dale in the 1930s, there continues to be no safety net for more recent immigrants, regardless of the extent of their own efforts, the overall economy, the unemployment rate, or the occurrence of circumstances beyond their control.[76]

[71] Interview, Peggy Fogelman, March 29, 2024.

[72] Interview, Dror (pseudonym), Spring 2024.

[73] Paul Lombardo, "Eugenics Laws Restricting Immigration," accessed April 13, 2024, http://www.eugenicsarchive.org/html/eugenics/essay9text.html

[74] Immigration and Nationality Act (INA) § 212(a)(4), 8 U.S.C. § 1182(a)(4).

[75] Personal Responsibility and Work Opportunity Reconciliation Act of 1996, Pub. L. No. 104–193, 110 Stat. 2105 (Aug. 22, 1996); and Illegal Immigration Reform and Immigrant Responsibility Act of 1996, enacted as Division C of the Defense Department Appropriations Act, 1997, Pub. L. No. 104–208, 110 Stat. 3008 (Sept. 30, 1996). See also Tanya Broder and Gabriel Lessard, "Overview of Immigrant Eligibility for Federal Programs," National Immigration Law Center, October 2023. Accessed April 13, 2024, https://www.nilc.org/issues/economic-support/overview-immeligfedprograms/

[76] Individuals fleeing persecution who have been granted asylum or refugee status are eligible for such benefits, but it has become increasingly difficult over the years for individuals to establish

References

Ackerman, Nathan and Jahoda, Marie. *Anti-Semitism and Emotional Disorder: A Psychoanalytic Interpretation.* New York: Harpers, 1950.
Ackerman, Natham and Jahoda, Marie. "The Dynamic Basis of Anti-Semitic Attitudes." *Psychoanalytic Quarterly* 17 (1948): 248–60.
Aldrich, C. Anderson "Preventive Medicine and Mongolism." *American Journal of Mental Deficiency* 52 (October 1947): 127–129.
American College of Obstetricians and Gynecologists, "Amniocentesis," April 2023. Accessed April 11, 2024, https://www.acog.org/womens-health/faqs/amniocentesis.
American Immigration Council, "Fact Sheet: Asylum in the United States," June 11, 2020. Accessed April 13, 2024, https://www.americanimmigrationcouncil.org/research/asylum-united-states.
American Psychiatric Association. *Diagnostic and Statistical Manual: Mental Disorders.* Washington, D.C.: American Psychiatric Association, 1952.
Anti-Defamation League, "Backgrounder: The Jewish Defense League," April 14, 2010. Accessed March 29, 2024, http://www.adl.org/extremism/jdl_chron.asp
Azad, Azade and Leander, Lina. "Children's Reporting about Sexual Versus Physical Abuse: Patterns of Reporting, Avoidance and Denial." *Psychiatry, Psychology, and Law* 22 (2015): 890–902.
Bakwin, Harry. "Informing the Parents of the Mentally Retarded Child." *Journal of Pediatrics* 49 (October 1956): 486–498.
Battegay, Caspar and Lubrich, Naomi. *Jewish Switzerland: 50 Objects Tell Their Stories/Jüdische Schweiz: 50 Objekteerzählen Geschichte.* Basel, Switzerland: Jüdisches Museum der Schweiz, Christoph Merian Verlag, 2018.
Bauman, Zygmunt. "Modernity, Racism, Extermination," in *Theories of Race and Racism: A Reader.* Edited by Les Black and John Solomos, 277–293. New York: Routledge, 2000.
Baumeister, Roy F., Dale, Karen, and Sommer, Kristin L. "Freudian Defense Mechanisms and Empirical Findings in Modern Social Psychology: Reaction Formation, Projection, Displacement, Undoing, Isolation, Sublimation, and Denial." *Journal of Personality* 66 (1998): 1081–1124.
Battista, Fabiana, Curci, Antonietta, Mangiulli, Ivan, and Ottgar, Henry. "What Can We Remember After Complex Denials? The Impact of False Denials on Memory." *Psychology, Crime and Law* 2021: https://doi.org/10.1080/1068316X.2020.1865956.
Beckman, L., Gustavson, K.-H. and Akesson, H.-O. "Studies of Some Morphological Traits in Mental Defectives: Mongolism and Unspecified Mental Deficiency." *Hereditas* 48, no. 12 (1962): 105–122.
Beddie, Alistair and Osmond, Humphry. "Mothers, Mongols, and Mores," *Canadian Medical Association Journal* 73, no. 3 (August 1955): 167–170.
Behrmann, Nicholas. "The Therapy Framework of Jewish 'Ethnotherapy'." Accessed January 4, 2024. https://www.linkedin.com/pulse/therapy-framework-jewish-ethnotherapy-behrmann-rabbi-pmp/.
Brockley, Janice. "Rearing the Child Who Never Grew. Ideologies of Parenting and Intellectual Disability in American History." In *Mental Retardation in America: A Historical Reader.*

such claims successfully. For basic information about asylum and the asylum application procedure, see American Immigration Council, "Fact Sheet: Asylum in the United States," June 11, 2020. Accessed April 13, 2024, https://www.americanimmigrationcouncil.org/research/asylum-united-states. A total of 17,700 people received asylum in fiscal year 2021, the fewest since 1994. Nicole Ward and Jeanne Batalova, "Refugees and Asylees in the United States," Migration Policy Institute, June 15, 2023, accessed April 13, 2024, https://www.migrationpolicy.org/article/refugees-and-asylees-united-states

Edited by Steven Noll and James W. Trent, Jr., 130–164. New York: New York University Press, 2004.
Broder, Tanya and Lessard, Gabriel. "Overview of Immigrant Eligibility for Federal Programs," National Immigration Law Center, October 2023. Accessed April 13, 2024, https://www.nilc.org/issues/economic-support/overview-immeligfedprograms/.
Buck, Pearl S. *The Child Who Never Grew*. New York: John Day, 1950.
Crookshank FG. *The Mongol in Our Midst: A Study of Man and His Three Faces*. New York: E.P. Dutton & Company, 1924.
Dean, John P. "Only Caucasian: A Study of Race Covenants." *Journal of Land and Public Utility Economics* 23, no. 4 (November 1947): 428–432.
Dobbs v. Jackson Health Organization, 597 U.S. __, 142 S.Ct, 2228 (2022).
Endelman, Todd M. "Jewish Self-Hatred in Britain and Germany," in *Two Nations: British and German Jews in Comparative Perspective*. Edited by Michael Brenner, Rainier Liedtke, and David Rechter, 331–363. Tubingen, Germany: Mohr Siebeck, 1999.
Faulkner, William. *Requiem for a Nun*, reprint ed. Knopf Doubleday Publishing Group, 2012.
Fiori, Katherine L., Considine, Nathan, and Merz, S. Eva-Maria. "Attachments, Social Network Size and Patterns of Social Exchange in Later Life." *Research on Aging* 33 (2011): 465–493.
Forman, Seth. *Blacks in the Jewish Mind: A Crisis of Liberalism*. New York: New York University Press, 2000.
Friedan, Betty. *The Feminist Mystique*. New York: Laurel Book, 1984.
Gierveld, Jeny De Jong. "Developing and Testing a Model of Loneliness." *Journal of Personality and Social Psychology* 53 (1987): 119–128.
Gilman, Sander L. *Jewish Self-Hatred: Anti-Semitism and the Hidden Language of the Jews*. Baltimore, MD: Johns Hopkins University Press, 1986.
Glazer, Nathan. "Negroes and Jews: The New Challenge to Pluralism." *Commentary* 38 (December 1964): 29–34.
Glenn, Susan A. The Vogue of Jewish Self-Hatred in Post World War II America." *Jewish Social Studies* 12, no. 3 (Spring-Summer 2006): 95–136.
Grebler, Anne Marie. "Parental Attitudes toward Mentally Retarded Children." *American Journal of Mental Deficiency* 55 (1951–1952): 475–483.
Grosberg, Michael. "From Feeble-Minded to Mentally Retarded: Child Protection and Changing Place of Disabled Children in Mid-Twentieth Century United States." *Paedagogica Historica* 47, no. 6 (2011): 729–747.
Halperin, Edward C. "Why Did the United States Medical School Admissions Quota for Jews End?" *American Journal of the Medical Sciences* 358, no. 5 (2019): 317–325.
Heinze, Andrew R. *Adapting to Abundance: Jewish Immigrants, Mass Consumption, and the Search for American Identity*. New York: Columbia University Press, 1990.
Huey, Edmund Burke. *Backward and Feeble-Minded Children*. Baltimore, MD: Warwick & York, 1912.
Illegal Immigration Reform and Immigrant Responsibility Act of 1996, enacted as Division C of the Defense Department Appropriations Act, 1997, Pub. L. No. 104–208, 110 Stat. 3008 (Sept. 30, 1996).
Immigration and Nationality Act (INA) § 212(a)(4), 8 U.S.C. § 1182(a)(4).
Kahane, Libby. *Rabbi Meir Kahane: His Life and Thought, Volume 1, 1932–1975*. CreateSpace Independent Publishing Platform, 2013.
Kaplan, Mordecai M. *The Future of the American Jew*. New York: Macmillan Company, 1948.
Kattsoff, L.O. "Editorial: Definition of a Situation: The Jew in America." *Social Forces* 27, no. 4 (May 1949): 461–464.
Kaufman, Gershen. *The Psychology of Shame: Theory and Treatment of Shame-Based Syndromes*. New York: Springer, 1989.
Kevles, Daniel J. "'Mongolian Imbecility': Race and Its Rejection in the Understanding of Mental Disease." In *Mental Retardation in America: A Historical Reader*. Edited by Steven Noll and James W. Trent, Jr., 120–129. New York: New York University Press, 2004.

Klein, Judith Weinstein. "Ethnotherapy with Jews." *International Journal of Mental Health* 5, no. 2 (Summer 1976): 26–38.

Klein, Kerwin Lee. "On the Emergence of Memory in Historical Discourse." *Representations* Winter 2000, no. 29 (Winter 2000): 127–150.

Knight, Amber, and Miller, Joshua. "Afterword: Prenatal Genetic Testing and Abortion after Dobbs," in *Prenatal Genetic Testing, Abortion, and Disability Justice*. Edited by Amber Knight and Joshua Miller, 193–198. New York: Oxford University Press, 2023.

Kosdon, Sari M., Keler, Alex, Berry, Jack W., Stiver, David J., Whiteford, Kelly, and O'Connor, Lynn E. "The Experience of Being Jewish: Living with Antisemitism." *PSOCIAl* 7, no. 1 (January–June 2021): ISSN:2422-619X.

Leeming, Dawn and Boyle, Mary. "Shame as a Social Phenomenon: A Critical Analysis of the Concept of Dispositional Shame." *Psychology and Psychotherapy: Theory, Research, and Practice* 77 (2004): 375–396.

Levinson, Abraham. *The Mentally Retarded Child*. New York: John Day, 1952.

Lewin, Kurt. "Everything Within Me Rebels: A Letter from Kurt Lewin to Wolfgang Kohler, 1933," Trans. Miriam Lewin and Gabriel Wickert, *Journal of Social Issues* 42 (1986): 39–47.

Lewin, Kurt. "Self-Hatred among Jews." *Contemporary Jewish Record* 4, no. 3 (June 1941): 219–232.

Lim, Polly QX, Shields, Nora, Nikolopoulos, Nikolaos, Barrett, Joanna T., Evans, Angela M., Taylor, Nicholas R. and Munteanu, Shannon E. "The Association of Foot Structure and Footwear Fit with Disability in Children and Adolescents with Down Syndrome." *Journal of Foot and Ankle Research* 8, no. 4 (2015) https://doi.org/10.1186/s13047-015-0062-0

Lindsay-Hartz, Janice. "Contrasting Experiences of Shame and Guilt." *American Behavioural Scientist* 27 (1984): 689–704.

Lombard, Paul. "Eugenics Laws Restricting Immigration. Accessed April 13, 2024, http://www.eugenicsarchive.org/html/eugenics/essay9text.html.

Mitchell, David and Snyder, Sharon. "The Eugenic Atlantic: Race, Disability, and the Making of an International Eugenic Science, 1800–1945." *Disability & Society* 18, no. 7 (2003): 843–864.

Nelson, William E. and Williams, Norman R. "Suburbanization and Market Failure: An Analysis of Government Policies Promoting Suburban Growth and Ethnic Assimilation." *Fordham Law Journal* 27, no. 1 (October 1999): 197–236.

Otgaar, Henry, Howe, Mark L., Memon, Amina, and Wang, Jianqing. "The Development of Differential Mnemonic Effects of False Denials and Forced Confabulations." *Behavioral Sciences and the Law* 32, no. 6 (2014): 718–731.

Otgaar, Henry, Howe, Mark L., Smeets, Tom, and Wang, Jianqing. "Denial-Induced Forgetting: False Denials Undermine Memory, But External Denials Undermine Belief." *Journal of Applied Research in Memory and Cognition* 5, no. 2 (2016): 168–175.

Otgaar, Henry, Romeo, Tameka, Ramakers, Nikki and Howe, Mark L. "Forgetting Having Denied: The 'Amnesic' Consequences of Denial." *Memory & Cognition* 46 (2018): 520–529.

Pelcovitz, Nathan A. "What About Jewish Anti-Semitism: A Prescription to Cure Self-Hatred." *Commentary* 3 (February 1947): 118–123.

Perlman, L. Daniel and Peplau, Anne. "Toward a Social Psychology of Loneliness," in *Personal Relationship 3: Personal Relationships in Disorder*. Edited by Robin Gilmour and Steve Duck, 31–43. London: Academic Press, 1981.

Personal Responsibility and Work Opportunity Reconciliation Act of 1996, Pub. L. No. 104–193, 110 Stat. 2105 (Aug. 22, 1996).

Picard, Jacques. *La Suisse et les juifs 1933–1945: Antisémitisme Suisse, defense du judäisme, politique internationaleenvers les emigrants et les refugiés*. Lausanne, Switzerland: Editions d'en bas, 2000.

Rautman, Arthur L. "Society's First Responsibility to the Mentally Retarded." *American Journal of Mental Deficiency* 54 (1949–1950): 155–162.

Reft, Ryan. "Segregation in the City of Angels: A 1939 Map of Housing Inequality in L.A." KCET website, January 19, 2017. Accessed January 15, 2024. https://www.kcet.org/shows/lost-la/segregation-in-the-city-ofangels-a-1939-map-of-housing-inequality-in-la.

Roe v. Wade, 410 U.S. 113, 93 S. Ct. 705, 35 L. Ed.2d 147 (1973).

Rogers, Dale E. *Angel Unaware: A Touching Story of Love and Loss*. Grand Rapids, MI: Revell, 1981 [1953].

Romeo, Tameka, Otgaar, Henry, Smeets, Tom, Lanström, Sara and Boerboom, Didi. "The Impact of Lying about a Traumatic Virtual Reality Experience in Memory." *Memory & Cognition* 47, no. 3 (2019): 485–495.

Satter, Beryl. *Family Properties: Race, Real Estate, and the Exploitation of Black Urban America*. New York: Macmillan, 2009.

Schmidt, Dennis Aldon. "Review of Mongolism with Special Emphasis Upon Theories of Its Etiology." M.D. thesis 2503, University of Nebraska Medical Center, 1960. Accessed February 19, 2024, https://digitalcommons.unmc.edu/mdtheses/2503

Selzer, Michael (Ed.). *"Kike!": A Documentary History of Anti-Semitism in America*. New York: World Publishing, 1972.

Shriver, Eunice Kennedy. "Hope for Retarded Children." *Saturday Evening Post* 235 (1952): 71–75.

Slaughter, Stella Stillson. *The Mentally Retarded Child and His Parent*. New York: Harper & Brothers, 1960.

Sokoloff, Leon. "The Rise and Decline of the Jewish Quota in Medical School Admissions." *Bulletin of the New YorkAcademy of Medicine* 68, no. 4 (November 1992): 497–518.

Spock, Benjamin. *Baby and Child Care*. New York: Pocket Books, 1957.

Stratthaus, Mary Ellen. "Flaw in the Jewel: Housing Discrimination against Jews in La Jolla, California." *American Jewish History* 84, no. 3 (September 1996): 188–219.

Teroni, Fabrice and Deona, Julien A. "Differentiating Shame from Guilt." *Consciousness and Cognition* 17, no. 3 (September 2008): 725–740.

The Arc. "About Us, Our History." Accessed February 19, 2024, https://thearc.orgabout-us/history/

Trent, James W. *Inventing the Feeble Mind: A History of Intellectual Disability in the United States*. New York: Oxford University Press, 2017.

United States Department of Agriculture, Economic Research Service, "Definitions of Food Security," October 25, 2023. Accessed March 29, 2024, https://www.ers.usda.gov/topics/food-nutrition-assistance/food-security-in-the-u-s/definitions-of-food-security/

United States Department of Justice, Federal Bureau of Investigation, "Terrorism 2000/2001," October 7, 2012. Accessed March 29, 2024, https://web.archive.org/web/20121007083445/https://www.fbi.gov/stats-services/publications/terror

Wald, N.J., Kennard, A., Hackshaw, A., & McGuire, A. "Antenatal Screening for Down Syndrome." *Journal of Medical Screening* 4, no. 4 (1997): 181–246.

Ward, Nicole and Batalova, Jeanne. "Refugees and Asylees in the United States," Migration Policy Institute, June 15, 2023. Accessed April 13, 2024, https://www.migrationpolicy.org/article/refugees-and-asylees-united-states.

"Where Down Syndrome Is Disappearing." CBS News, August 15, 2017. Accessed February 19, 2024, https://www.cbsnews.com/news/down-syndrome-iceland/.

Zenderland, Leila. *Measuring Minds: Henry Herbert Goddard and the Origins of American Intelligence Testing*. New York: Cambridge University Press, 1998.

Lectures

Hirsch, Marianne. "The Generation of Postmemory: What, Who, When, Where, How?" Sponsored by the Museum of Jewish Heritage, New York, NY. January 9, 16, 23, 2023 [virtual].

Interviews

Dror (pseudonym), Winter 2023; Spring 2024.
Dale Zand, October 7, 2023.
Peggy Fogelman, March 28, 2024.

Index

A
Abandonment, 45, 50, 51, 76
Adverse childhood experience (ACE), 48
Anti-miscegenation, 14
Antisemitism, 8, 11, 14, 26, 49, 71–78
Aryan, 18, 27, 73
Assimilation, 59

B
Better baby contests, 24
Bronx, 2, 29, 53, 54, 59
Brooklyn, 53, 55, 59, 79
Buck v. Bell, 24, 25

C
Cholera, 5, 12, 18
Confessional transfer, 9
Conscription, 8, 11
Consumer attitudes, 79
Coughlin, C. Father, 25, 29, 30

D
Darwin, C., 5, 7, 15
Davenport, C., 15
Denials, 77
Desertions, 51, 52
Down syndrome, 1, 2, 66, 68, 69
Druya, 3, 5, 8, 12, 30

E
Economic burden, 28
Economic difficulties, 49
Economic factors, 7
Economic opportunities, 77
Economic restrictions, 22
Economic situation, 73
Economic thought, 15
Ellis Island, 16, 43
Emergency Quota Act of 1921, 41
Employment, 6, 42, 51, 56, 58, 59, 77
English, 22, 23, 39, 71
Eugenics, 1–30, 40–60, 66–78
Eugenics Research Association, 14, 21
Exclusion, 25, 59, 78–81

F
Feeble-minded, 10, 16, 67
Florinsky, V., 3, 5
Food insecurity, 80

G
Gamalei, N., 6
Germany, 11, 25, 29, 30, 53, 71, 75
Goddard, H.H., 16
Grant, M., 15, 19
Great Depression, 56, 58, 59
Guilt, 24, 47, 59, 69, 71, 76, 81

S. Loue, *From Public Policy to Family Dynamics*, SpringerBriefs in Social Policy, https://doi.org/10.1007/978-3-031-71818-2

H
Hall, P.F., 19
Handicapped, 27, 67

I
Identity, 8, 9, 22, 43–45, 50, 51, 65–81
Immigration Act of 1924, 41
Immigration policy, 40–54, 81
Industrialization, 15
International Ladies' Garment Workers' Union (ILGWU), 43
Isolation, 39, 77–81
Iudin, T., 7

J
Jewish self-hate/self-hatred, 71–78
Judaism, 7, 9, 26

K
Koltsov, N.K., 9
Kropotkin, P., 9

L
Laughlin, H.H., 14, 20, 21, 25
Literacy, 22
Lowicz, 40, 41, 44, 45, 49, 59

M
Medicine, 6, 28, 29, 66, 76
"Melting pot", 23
Memories, 30, 53, 59, 60, 65, 66, 77
Mental disability, 69
Mental distress, 78
Mental health professional, 75
Mental illness, 7, 17, 18, 28
Mental retardation, 2, 67, 70
Mother, 24, 39, 46–48, 67, 71–73, 76, 78, 79

N
Nazis, 2, 9, 11, 18, 20, 21, 25–30, 72, 74, 78
Nuremberg Laws, 21

P
Pale of Settlement, 3–5, 12, 13, 30
Physicians, 1, 6, 7, 15, 16, 27, 29, 66, 67, 80
Plague, 5
Pogroms, 12, 13, 26, 30
Poland, 5, 12, 29, 39, 43, 46, 47, 53, 54, 79
Poverty, 3, 55–60, 76, 78–81
Power, 10, 19, 77
Psychohistory, vii
Psychosis, 7, 17
Psychosocial crisis, 44
Psychosocial growth, 44
Public health, 6, 16, 24

Q
Quarantine, 18

R
Races, 3, 5, 7, 9, 11, 15, 16, 18–23, 25–29, 45, 66, 73
Racial hygiene, 26–29
Ross, E.A., 16
Russia, 3, 5, 9, 10, 12, 14, 16, 19, 22, 29, 50, 53, 54
Russian Eugenics Society, 9, 11

S
Schlovskii, I., 9
Segregation, 7, 9, 14, 73
Self-hate, 75
Shame, 22, 49, 50, 60, 67, 69, 71, 74, 76–81
Sterilizations, 7, 9, 14, 15, 24, 27, 28, 67
Switzerland, 67, 71, 73–76

T
The Arc, 69
Trauma, 48, 65, 77
Typhoid, 5
Typhus, 5, 12, 18

U
Union, 8, 11, 43, 58

W
Welfare policy, 81
Work, 5, 20, 23, 51, 53, 56, 58, 81

SPRINGER NATURE

GPSR Compliance

The European Union's (EU) General Product Safety Regulation (GPSR) is a set of rules that requires consumer products to be safe and our obligations to ensure this.

If you have any concerns about our products, you can contact us on ProductSafety@springernature.com

In case Publisher is established outside the EU, the EU authorized representative is:

Springer Nature Customer Service Center GmbH
Europaplatz 3
69115 Heidelberg, Germany

The manufacturer's authorised representative in the EU is Springer Nature Customer Service Centre GmbH, Europaplatz 3, 69115 Heidelberg, Germany. If you have any concerns regarding our products, please contact ProductSafety@springernature.com

Printed and bound by CPI Group (UK) Ltd, Croydon, CR0 4YY
01/10/2026
02225880-0008